T0291123

Mitigation and Adaptation to Climate Change

Mitigation and Adaptation to Climate Change

Edited by Dale Sullivan

SYRAWOOD
PUBLISHING HOUSE

New York

Published by Syrawood Publishing House,
750 Third Avenue, 9th Floor,
New York, NY 10017, USA
www.syrawoodpublishinghouse.com

Mitigation and Adaptation to Climate Change
Edited by Dale Sullivan

International Standard Book Number: 978-1-64740-342-3 (Hardback)

Cataloging-in-publication Data

Mitigation and adaptation to climate change / edited by Dale Sullivan.
    p. cm.
Includes bibliographical references and index.
ISBN 978-1-64740-342-3
1. Climate change mitigation. 2. Climatic changes. 3. Environmental protection. Adaptation (Biology) .
I. Sullivan, Dale.
TD171.75 .M58 2023
363.738 746--dc23

Table of Contents

Preface VII

Chapter 1 **Food Security Concerns, Climate Change and Sea Level Rise in Coastal Cameroon**...1
Wilfred A. Abia, Comfort A. Onya, Conalius E. Shum,
Williette E. Amba, Kareen L. Niba and Eucharia A. Abia

Chapter 2 **Maize, Cassava and Sweet Potato Yield on Monthly Climate**.......................14
Floney P. Kawaye and Michael F. Hutchinson

Chapter 3 **Risks of Indoor Overheating in Low-Cost Dwellings on the South African Lowveld**..35
Newton R. Matandirotya, Dirk P. Cilliers, Roelof P. Burger,
Christian Pauw and Stuart J. Piketh

Chapter 4 **Triple Helix as a Strategic Tool to Fast-Track Climate Change Adaptation in Rural Kenya: Case Study of Marsabit County**....................53
Izael da Silva, Daniele Bricca, Andrea Micangeli,
Davide Fioriti and Paolo Cherubini

Chapter 5 **Climate Change Adaptation through Sustainable Water Resources Management in Kenya: Challenges and Opportunities**............................76
Shilpa Muliyil Asokan, Joy Obando, Brian Felix Kwena and
Cush Ngonzo Luwesi

Chapter 6 **Climate Change Impact on Soil Moisture Variability: Health Effects of Radon Flux Density within Ogbomoso**......................................87
Olukunle Olaonipekun Oladapo, Leonard Kofitse Amekudzi,
Olatunde Micheal Oni, Abraham Adewale Aremu and
Marian Amoakowaah Osei

Chapter 7 **Global Strategy, Local Action with Biogas Production for Rural Energy Climate Change Impact Reduction**..103
A. S. Momodu, E. F. Aransiola, T. D. Adepoju and I. D. Okunade

Chapter 8 **Climate Change Implications and Mitigation in a Hyperarid Country: A Case of Namibia**...122
Hupenyu A. Mupambwa, Martha K. Hausiku,
Andreas S. Namwoonde, Gadaffi M. Liswaniso,
Mayday Haulofu and Samuel K. Mafwila

Chapter 9   **GIS-Based Assessment of Solar Energy Harvesting Sites and
              Electricity Generation Potential**...**144**
              Mabvuto Mwanza and Koray Ulgen

Chapter 10   **Climate Change Adaptation in Southern Africa: Universalistic
              Science or Indigenous Knowledge or Hybrid**...**192**
              Tafadzwa Mutambisi, Nelson Chanza, Abraham R. Matamanda,
              Roseline Ncube and Innocent Chirisa

Chapter 11   **Farmers' Adaptive Capacity to Climate Change
              in Africa: Small-Scale Farmers**...**207**
              Nyong Princely Awazi, Martin Ngankam Tchamba,
              Lucie Felicite Temgoua and Marie-Louise Tientcheu-Avana

              Permissions

              List of Contributors

              Index

PREFACE

Over the recent decade, advancements and applications have progressed exponentially. This has led to the increased interest in this field and projects are being conducted to enhance knowledge. The main objective of this book is to present some of the critical challenges and provide insights into possible solutions. This book will answer the varied questions that arise in the field and also provide an increased scope for furthering studies.

Climate change can be countered by two strategies, which include mitigation and adaptation. Mitigation addresses the causes of climate change, whereas adaptation addresses the consequences. Adaptation is the process of making changes in response to climate change. Mitigation is human interference for decreasing the sources of greenhouse gas emissions, to prevent a rise in temperatures. Adaptation focuses on seeking appropriate measures to prevent the impact of climate change. It is quite difficult to forecast adaptation solutions since they differ from place to place and consist of several trade-offs. The first stage in adapting climate change is to understand the local risks and devise strategies for managing them. The next step is to implement systems to respond to the effects being experienced in the present while preparing for the uncertainty in the future. This book provides comprehensive insights on climate mitigation and adaptation. It strives to provide a fair idea about climate change and to help develop a better understanding of the latest advances in its management. This book is a resource guide for experts as well as students.

I hope that this book, with its visionary approach, will be a valuable addition and will promote interest among readers. Each of the authors has provided their extraordinary competence in their specific fields by providing different perspectives as they come from diverse nations and regions. I thank them for their contributions.

Editor

1

Food Security Concerns, Climate Change and Sea Level Rise in Coastal Cameroon

Wilfred A. Abia, Comfort A. Onya, Conalius E. Shum, Williette E. Amba, Kareen L. Niba and Eucharia A. Abia

Contents

Introduction .
Highlights of Coastal Cameroon
Coastal Zones and Risk Factors of Agricultural Productivity

W. A. Abia (✉)
Laboratory of Pharmacology and Toxicology, Department of Biochemistry, Faculty of Science, University of Yaounde 1, Yaounde, Cameroon

School of Agriculture, Environmental Sciences, and Risk Assessment, College of Science, Engineering and Technology (COSET), Institute for Management and Professional Training (IMPT), Yaounde, Cameroon

Integrated Health for All Foundation (IHAF), Yaounde, Cameroon

C. A. Onya
Natural Resources and Environmental Management, University of Buea, Buea, Cameroon

C. E. Shum · W. E. Amba
School of Agriculture, Environmental Sciences, and Risk Assessment, College of Science, Engineering and Technology (COSET), Institute for Management and Professional Training (IMPT), Yaounde, Cameroon

K. L. Niba
School of Agriculture, Environmental Sciences, and Risk Assessment, College of Science, Engineering and Technology (COSET), Institute for Management and Professional Training (IMPT), Yaounde, Cameroon

Integrated Health for All Foundation (IHAF), Yaounde, Cameroon

E. A. Abia
Integrated Health for All Foundation (IHAF), Yaounde, Cameroon

Climate Change and Sea Level Rise (CC/SLR) and Food Security in Coastal Cameroon
   Climate Change and Coastal Areas
   Sea Level Rise (SLR)
   Climate Change and Sea Level Rise (CC/SLR)
Conclusions and Future Prospects
References

Abstract

Food security is a major public health priority in Cameroon, amidst climate change and sea level rise (CC/SLR), vis-à-vis the ever-increasing population growth with associated challenges. CC/SLR, singly or combine, is well known to have severe impacts on agricultural productivity, food security, socioeconomic activities and ecosystem (environment, plant and animal) health systems in coastal areas. They contribute to natural disasters including erosion, flooding, inundation of coastal lowlands, and saltwater intrusion, altogether reducing agricultural productivity. Additionally, these disasters provoke adverse animal, human, and environmental health implications; reduction in tourism; and potential close of some socioeconomic activities that constitute secondary (after agriculture), or main source of livelihood/income for many coastal indigents. Although there are inadequate reports on the impacts of CC/SLR, preliminary reports point to negative effects on crop production and socioeconomic activities in coastal Cameroon. This chapter highlights the susceptibility of coastal Cameroon agriculture and socioeconomic activities to CC/SLR. Furthermore, it has propose agricultural (CC/SLR and non-climatic) and educational intervention socioeconomic strategies for the mitigation and adaptation to CC/SLR and for sustainable agricultural productivity in coastal Cameroon. The proposed strategies may provide a small contribution toward a wider multi-stakeholder pool of strategies and which, when applied, may enhance food security in coastal Cameroon amidst CC/SLR and promote socioeconomic and touristic activities while reducing negative implications on animal, plant, human, and environmental health.

Keywords

Agriculture · Climate change · Sea level rise · Food security · Ecosystem health

Introduction

Food security is a major concern in feeding the world's estimated 9.8 billion people by 2050 (Worldometer 2020). The demand for food by 2050 will increase by an estimated 60%. This will be associated with broader economic and societal issues. Thus, there is increasing need for sustainable agriculture (Breene 2016; Abia et al. 2016; Center for Development Research (ZEF), Forum for Agricultural Research in Africa (FARA), Institute of Agricultural Research for Development (IRAD) 2017) toward sustainable food supply. Food security exists when all people, at all times, have physical and economic access to sufficient safe and nutritious food that meets their dietary needs and food preferences for an active and healthy life (Food and Agriculture Organization (FAO) 2010). In addition, food security has three dimensions. These include (i) availability of food (which consist of three elements related to production, allocation, and exchange); (ii) access to food (that is connected with affordability, e.g., income and wealth, provision, and preferences); and (iii) utilization of food (focusing on the essential elements that are associated with dietetic and social values as well as food safety issues). Additionally, it is vital to ensure that the available and accessible food is safe. Furthermore, food waste should be minimized as much as possible (zero food waste concept. According to the Codex Alimentarius Commission (CAC 2003), food safety is the assurance that food will not cause harm to the consumer when it is prepared and/or eaten according to its intended use.

Food security is generally affected by several factors. Climate change remains one of the most devastating factors. Notwithstanding, both climatic and non-climatic factors, singly and collectively, hinder agricultural productivity. These factors include increase in temperature, fluctuation in rainfall (periods and amounts) and population growth, and sea level rise. Their effects on coastal agricultural productivity and sustainability may be a concern (Schiermeier 2018). They may provoke inundation, soil erosion, and saltwater intrusion (Gopalakrishnan et al. 2019).

In Cameroon, agriculture and food associated sectors provides employment to an estimated 75% of the adult-working population (mainly small-scale peasant farmers (NIS 2010), contributing 30% to the gross domestic product (GDP) and account for approximately half of total earnings from exports (DSCN 2002). The sustainability of the agricultural sector in Cameroon is essential (Abia et al. 2016) and sustaining the food sector (ZEF, FARA, IRAD 2017). However, there is inadequate focus on the limitations to the country's coastal agricultural productivity. Coastal Cameroon's agricultural productivity is likely already and may continue to experience adverse impacts of climate change and sea level rise (CC/SLR) especially in terms of area of inundation, soil erosion, flooding, salinity intrusion, and reduction in crop production. This may have serious repercussions on farmers, food security and safety, as well as on the ecosystem (plant, animal and human) health. An early awareness and preparation toward helping coastal farmers cope with CC/SLR is relevant. This chapter highlights the vulnerabilities of coastal Cameroon's agricultural productivity and ecosystem health to CC/SLR induced hazards, with proposed mitigation strategies.

Highlights of Coastal Cameroon

The coastal lowlands of Cameroon is located between the Atlantic Ocean and the western highlands in the northeast and the south Cameroon plateau in the southeast and covers 402 km of coastline. The coastal Cameroon spans 15–150 km inland from the Gulf of Guinea with an average elevation of 90 m. The coastal zone of Cameroon has three sedimentary basins (Campo Kribi, Douala, and Rio-del Rey). These are known to be potentially rich in hydrocarbons and are currently being exploited by petroleum companies. The northern part of the coast (including Idenau, Debundscha, Batoke, Bota, and Down Beach) is characterized by a small population size and very few industries and suffering impacts from the volcanic eruptions of Mount Cameroon. For example, during eruption, lava flows obstruct road networks, destroying crops, and induce rise in ocean water thereby killing fishes and other marine ecosystems. The coastal zone harbors the coldest place in Cameroon, Debundscha, which is at the foot of the Mount Cameroon, which experience the highest rainfall (annual average: 11,000 mm). The coastal region is characterized by equatorial climate with less dry (~3 months) and wet (~9 months) seasons alternating. The coastal line has high humidity mainly associated with the Guinea monsoon winds. The center of the coast, i.e., Douala, has an estuarine system of river Wouri and is the part with highest human/anthropogenic activities. Additionally, the central coast has the highest coastal population size and is home to approximately 60% of Cameroon's industries (Alemagi et al. 2006). It harbors the countries important industrial and environmental interests (Onguéné et al. 2015). The southern coast area (Kribi) harbors the smallest coastal population size and has few industries. The characteristics of coastal Cameroon have been presented variedly e.g., based on water, salt, and nutrient budget of the two estuaries (Gabche and Smith 2002) and based on characteristics of coastal vulnerability to climate change (Leal Filho et al. 2018).

Coastal Zones and Risk Factors of Agricultural Productivity

Coastal areas are likely to be more vulnerable to climate change than inland areas because, in addition to changes in flooding, temperature, and precipitation, coastal lowlands are frequently affected by sea level rise (SLR) and sea wave heights. Increasing Greenhouse Gas (GHG) emissions may raise the average atmospheric temperature by 1.1 °C to 6.4 °C over the next century, with possible thermal expansion of the oceans, rapid melting of ice sheets, and consequently SLR (Intergovernmental Panel on Climate Change (IPCC) 2007a). On the average, the global SLR stood at the rate of 1.8 mm per year from 1961 to 1993 and at the rate of about 3.1 mm per year from 1993 to 2003 (IPCC 2007b). Even if GHG emissions were stabilized soon, thermal expansion and deglaciation would continue to trigger SLR for many decades. Furthermore, the continuous growth of GHG emissions and associated global warming could well promote SLR of 1–3 m in this century, and unexpectedly rapid breakup of the Greenland and West Antarctic ice sheets might produce a 5 m SLR (Church et al. 2001) and may rise to 7.5 m by 2020 (Bamber et al. 2019). Altogether, the IPCC Third Assessment Report of 2001 projected a global average SLR of between 20 and 70 cm between 1990 and 2100 using the full range of IPCC GHG scenarios and a range of climate models (IPCC 2001). Recently, SLR projection until 2030 was reported in the "Special IPCC Report on the Ocean and Cryosphere in a Changing Climate" (IPCC 2019). In the report, two scenarios for GHG emissions are considered: a "low" scenario (known as RCP2.6), with strong reduction of global greenhouse gas emission, such that global warming will probably not exceed 2 °C, and a "high" scenario (referred to as RCP8.5), in which no measures are taken to limit

GHG emissions. Altogether, it is assumed that the high scenario may lead to SLR of up to 5 m of the global average sea level in 2030 (IPCC 2019).

Coastal areas are generally vulnerable to anthropogenic influences such as dense population, industrialization, and agricultural activities (Amosu et al. 2012). In the west and central African sub-regions, erosion of beaches is among the major ecological problems (Ibe and Awosika 1991). Due to coastal erosion and SLR, the surface area of the coastal administrative capital of the Gambia, Banjul, may disappear within approximately 50–60 years, thereby jeopardizing livelihood for over 42,000 people (Jallow et al. 1999). SLR causes devastating effects, which could include loss of land, population displacement, loss of economic gain, loss of urban infrastructures and amenities, submersion of agricultural lands, wetlands (or biodiversity) loss, and even the disruption of several ecosystems (Dasgupta et al. 2009).

Climate Change and Sea Level Rise (CC/SLR) and Food Security in Coastal Cameroon

Combined climate change (e.g., shifting weather patterns) and sea level rise (e.g., increase the risk of catastrophic flooding) (CC/SLR) has continued to threaten global agricultural production, socioeconomic activities and planetary health in an unprecedented manner for a while now, despite continuous efforts world over to mitigate it. There is urgent need for drastic actions, now more than ever; otherwise it will be more complex and expensive adapting to impacts of CC/SLR in the future (United Nations 2019).

SLR is generally referred to as "an increase in the level of the world's oceans due to the effects of global warming." Basically, a warming climate may cause seawater to expand and ice over land to melt. Both scenarios in combination may cause sea levels to rise (SLR). Thus, SLR is one of the major effect of climate change (CC), with rising waters threatening to inundate small-island nations and **coastal** regions in various parts of the world (Mimura 2013) and Cameroon (Fonteh et al. 2009). The effects of climate change, SLR, and both CC/SLR are discussed below.

Climate Change and Coastal Areas

Among all the environmental challenges known to have overwhelmed the planet since the 1980s, it has been estimated that more than 70% of them are linked to climate change (Lambi and Kometa 2014). In the wake of natural disaster such as droughts, SLR, floods, tropical depressions as hurricanes, storms, and heat waves, there has been an overwhelming negative impact on humankind, the environment, and economic livelihoods (Living with Risk 2002; Associated Program on Flood Management 2009; Brown et al. 2013). Cameroon is exposed to the impacts of climate change particularly her territories located in the Sahelian zone (which are extremely threatened by effects of desertification) and coastal areas (that are highly vulnerable to SLR) (Banseka and Levesque 2018). Partly due to the impacts of

CC/SLR, coastal Cameroon is already facing extreme weather phenomena such as heavy rainfall, violent winds, high temperatures, and drought, which endanger communities' ecosystems and the services they provide (Molua 2006; Fonteh et al. 2009; Banseka and Levesque 2018).

The major risks of climate change are inundation, soil erosion, and saltwater intrusion (Gopalakrishnan et al. 2019) and which negatively affects agricultural productivity, with the worst impact in the coastal areas. Additionally, it is speculated that by 2080, coastal West Africa may experience a high-risk level of flooding provoked by climate change (Nicholls and Tol 2006).

In addition to the social and human costs, the economic cost of the impacts of climate change is immense. This includes decrease or losses in agricultural productivity due to droughts and increased variability of rainfall due to increased numbers and intensity of natural disasters such as SLR (Banseka and Levesque 2018). It is speculated that, if nothing is done to address climate change, the "cost of inaction" may be huge and is estimated to be between 5% and 20% of world GDP, whereas the cost of "acting" is estimated at only 1% or 2% (Stern 2006).

In order to contain the devastating effects and or influences of climate change on the environment, agricultural productivity, and ecosystem (animal, human, plant, environmental) health, which has jeopardized the entire planetary systems, two types of policy response measures are needed: mitigation (efforts to limit GHG emission) and adaptation (actions taken to reduce the negative consequences of changes in the climate).

Adaptation capacity designates the ability for society to plan for and respond to change in a way that makes it better equipped to manage its exposure and sensitivity to climate change. Nearly 2.4 billion people (about 40% of the world's population), live within 100 km on coastal strip around the globe, making a total of 60 miles of total land surface occupation per/inhabitant of the coastal strip. Thus, oceans coastal and marine resources are very important for people living in coastal communities, which represent 37% of global population (United Nations Factsheet 2017). The term coastal zone is a region where there is interaction of the sea and land processes, for example, the city of Limbe in the coastal Cameroon.

Globally, the most common adaptation and mitigation measures used is the "Ecosystems Based Adaptation" (EBA) approach. This involves the conservation, sustainable management, and restoration of ecosystems to adapt to the advert effects of climate change (Convention on Biological Diversity (CBD) 2009, 2018). This approach will help people to take into account, manage ecosystems in ways that permit them to adapt to climate change in coastal areas. The United Nations Environmental Program (UNEP) has laid-down foundation guides for EBA options under the UNEP building capacity for coastal EBA for small islands Development States (project funded by European Commission). This guide is a strategic resource geared at helping environmental and adaptation managers and planners, mainly in governmental departments and civil society organizations. It facilitates baselines knowledge and built broad understanding of the principles and concepts of coastal EBA (UNEP 2016).

In synopsis, "EBA implement and support, environmental decision makers in choosing, implementing, monitoring, evaluating and over time adaptively managing coastal areas." Mangroves, coral reef, estuaries, seagrass beds, dune communities, and other systems on or near shorelines do serve critical ecological functions which are beneficial to human society. Some of these functions include fisheries, storm protection, floods mitigation, erosion control water storage, ground water recharge, pollution abatement, retention, and cycling of nutrients as well as sediments. In a similar manner, the Convention on Biological Diversity (CBA) has also acknowledged and recognized the potential importance of EBA in meeting this challenge (CBD 2009, 2018).

Around the globe, the coastal zone management Act (CZMA) introduced in 1972 is equally applicable. This sought to balance economic development with environmental conservation, mainly by avoiding the scenario of specifying a defensive definition approach to climate change management. The National as well as international CZMA programs encourage various countries of the globe to develop and implement CZMA plans to protect, restore, and develop the resources of their national coastal zones for present and future generations. A good number of states still recognize the importance pre-emptive action to address their vulnerability to climate change (C2ES 2011). Some mitigation innovations include beach nourishment, coastal fortification, and a reactionary approach which includes seawalls, groin and jetty construction, and inshore artificial reefs.

In Cameroon, the reality of climate change is widely acknowledged. It is the consequence of increasing temperatures caused by atmospheric GHGs, altering the functioning of the ecosystems. According to the Fourth Assessment Report of the **IPCC**, the efficiency is difficult to assess because of natural adaptation and non-climatic factors (IPCC 2007c). Furthermore, 70% of GHG emissions observed between 1970 and 2004 was caused by human activities (IPCC 2007a). The IPCC Synthesis Report suggests that a continuation of the present policies to mitigate climate change would probably lead GHG emissions to further in the coming decades (IPCC 2007d). Therefore, for improved and sustainable agricultural production, there is a need for continuous monitoring and forecasting and use of crops and varieties that are more resistant to drought and adaptation of suitable planting methods (Molua 2006). Furthermore, there is need for expansion of farm size, livelihood diversification, and usage of organic fertilizers as potential adaptation options (Epule and Bryant 2016). However, non-climatic factors such as deforestation, poor governance, inadequate access to farm inputs (e.g., fertilizers, increased economic opportunities elsewhere and a breakdown of cultural practices) cannot be minimized (Epule and Bryant 2016).

Sea Level Rise (SLR)

The impact of sea level rise (SLR) on developing countries is overwhelming (Dasgupta et al. 2009). The effects will likely be more in coastal lowlands. SLR is expected to pose unique challenges partly due to the resultant saline contamination.

SLR may provoke more salt in soils and too much salt in soil can ruin crop yield (e.g., through restriction of water and nutrient uptake by the plant) and render farmlands or fields useless (Schiermeier 2018). SLR in coastal zones could potentially lead to land loss through inundation, erosion of coastal lands, increased frequency and extent of storm-related flooding, and increased salinity in estuaries and coastal freshwater aquifers (Gopalakrishnan et al. 2019).

Fonteh et al. (2009) have revealed possible implications of future SLR on the ecosystems and economic activities along the coast of Cameroon using mapping and valuation approaches. This was partly associated with the high ecological and economic value of the area. It was speculated that an estimated 112–1216 km^2 (1.2–12.6%) of the coastal area is likely to be lost from a 2–10 m (equivalent to a low scenario by 2050 and high scenario by 2100) flooding, respectively. Furthermore, approximately between 0.3% and 6.3% of ecosystems (estimated to worth US$ 12.13 billion/year) could be at risk of flooding by the years 2050 and 2100. The areas under a serious threat contain mangroves, sea and airport, residential and industrial areas, and to a lesser extent, main plantation crops of banana and palms (Fonteh et al. 2009). Although there is inadequate information on the consequences, it may be speculated that this may have adverse effects on coastal agriculture production and may be a threat to food security and safety, as well as on the socio-economy of the coastal Cameroon. Wetland losses and loss of productive mangrove ecosystem will also occur with a SLR. According to the TOPEX/Poseidon and Jason satellite data, the rate of SLR in Cameroon is 2–2.4 mm/year between 1993 and 2004 (NASA 2008). Thus, the low-lying coastal areas are physically and socioeconomically vulnerable to impacts of SLR. There is need to take prompt actions toward mitigating the effects of SLR provoked natural disasters in coastal Cameroon (Fonteh et al. 2009).

Climate Change and Sea Level Rise (CC/SLR)

Climate change and sea level rise (CC/SLR) are not new concepts, even though their synergistic effects seem minimized and less talked about vis-à-vis climate change or food security alone or in combination. SLR is a direct consequence of global climate change. It appears to get worst as population growth increases (Gommes and du Guerny 1998). The CC/SLR constitute a major hindrance to agricultural productivity (Schiermeier 2018; Ogbuabor and Egwuchukwu 2017) and may exert adverse effects on ecosystem health (Nicholls et al. 2011) in coastal areas.

As climate change continues to provoke SLR in coastal Cameroon, inundation of low-lying coastal areas increases continuously and with saltwater, and gradually contaminating the soil. Although rainfall can dissipate these salts, climate change is also increasing the frequency and severity of extreme weather events, including droughts and heat waves. This leads to intensive use of groundwater for drinking and irrigation, which further depletes the water table and allows salt to leach into soil. In some parts of the world, especially low-lying river deltas, local land is sinking (known as subsidence), making sea levels that much higher (Nicholls et al. 2007).

Thus, even without climate change, coastal areas such as coastal Cameroon would still experience slow relative SLR due to these non-climatic processes and hence increased flooding and damage cost through time (Nicholls 2002; Nicholls et al. 2007, 2011).

In addition to adverse impact on agricultural productivity, CC/SLR constitute a nuisance in coastal ecosystem health and socioeconomic activities. This is particularly worse in the developing countries with inadequate capacity to manage associated repercussions on land use, populations' evacuation/displacement and juxtaposition of livelihood sources, i.e., agriculture and touristic activities in the case of coastal Cameroon and elsewhere. For example, in coastal Cameroon, CC/SLR may lead to flooding which may provoke internal displacements with associated joblessness. The coastal Cameroon's situation is made worse by the eruption of Mt. Cameroon, which releases lava flow that further reduce agricultural land and destroy crops and marine lives whenever it erupts – a scenario which is arguably by the assumed post eruption increase in soil fertility. Apparently, coastal zones in different geographic areas, with varied anthropogenic activities, non-similar efforts against SLR, and varied effectiveness of erosion-driving forces such as waves and tides, are expected to react differently to SLR. Thus, considering the CC/SLR projected likelihood of high-risk flooding in the lowlands of coastal West Africa (including Cameroon) by 2080 (Nicholls and Tol 2006), there is a need for constant monitoring and timely mitigating actions to salvage coastal Cameroon from the natural disaster. To this effect, we speculate that relevant efforts to mitigate and adapt to CC/SLR may include intensive farmers' education on good agricultural practices, creation of nurseries for improved and climate smart food varieties, irrigation and vertical farming, and a shift from subsistence-to-industrial farming. Application of these strategies may enhance food security, ecosystem health and socioeconomic and touristic activities in the coastal Cameroon.

Conclusions and Future Prospects

Food security is a major public health priority in Cameroon, vis-à-vis the ever-increasing population growth with associated challenges. Risk factors to agricultural productivity such as combined climate change and sea level rise (CC/SLR) require attention, particularly in coastal Cameroon. CC/SLR have negative impacts on agricultural productivity and socioeconomic activities in coastal morphology. This is through its contributions to natural disasters including erosion, flooding, inundation of coastal lowlands, and saltwater intrusion, which all have a significant impact on crop productivity and output. Socioeconomically, these disasters may provoke adverse animal, human, and environmental health implications; reduction in tourism; and potential close of some socioeconomic activities (e.g., beer parlors, and small roadside restaurants) that constitute as secondary (after agriculture), or main source of livelihood/income for many coastal indigents.

Sustainable agricultural productivity in Cameroon is essential, moreover in coastal Cameroon amidst CC/SLR. This requires concerted actions of stakeholders' (government, local civil society organizations, individual, and common initiative farmers groups, and international bodies). Such efforts may propose and develop sustainable strategies toward adaptation and mitigation of risk factors of agricultural production and socioeconomic growth in coastal Cameroon. Potential adaptation options may span from agricultural (CC/SLR and non-climatic) to educational intervention socioeconomic strategies. CC/SLR strategies include expansion of farm size, usage of organic fertilizers, creation of nurseries for improved and climate smart crop varieties, irrigation and vertical farming, and a shift from subsistence-to-industrial farming. Non-climatic strategies may include afforestation, good governance, ensuring adequate access to farm inputs (e.g., fertilizers), promotion, and increase of local agribusiness opportunities. Educational intervention socioeconomic strategies may include intensive farmers' education on good agricultural practices, and the diversification of livelihood. These strategies may provide a small contribution toward a wider multi-stakeholder pool of strategies and which, when applied, may enhance food security in coastal Cameroon amidst CC/SLR and promote socioeconomic and touristic activities while reducing negative implications on animal, plant, human, and environmental health.

References

Abia WA, Shum CE, Fomboh RN, Ntungwe EN, Ageh MT (2016) Agriculture in Cameroon: proposed strategies to sustain productivity. Int J Res Agric Food Sci 2(2):1–12

Alemagi D, Oben PM, Ertel J (2006) Mitigating industrial pollution along the Atlantic coast of Cameroon: an overview of government efforts. Environmentalist 26:41–50

Amosu AO, Bashorun OW, Babalola OO, Olowu RA, Togunde KA (2012) Impact of climate change and anthropogenic activities on renewable coastal resources and biodiversity in Nigeria. J Ecol Nat Environ 4(8):201–211

Associated Program on Flood Management (2009) Integrated flood management concept paper. Flood management policy series (World Meteorological Organization, WMO) No. 1047. Available online at: http://www.apfm.info/pdf/concept_paper_e.pdf. Accessed 20 June 2020

Bamber JL, Oppenheimer M, Kopp RE, Aspinall WP, Cooke RM (2019) Ice sheet contributions to future sea-level rise from structured expert judgment. PNAS 116(23):11195–11200. https://doi.org/10.1073/pnas.1817205116

Banseka H, Levesque LCT (2018) Cameroon: preparing the national adaptation plan for climate change (NAPCC) and its investment strategy (#492). In Gunya KJ, Philip R, Sara Oppenheimer S (eds) GWP global water partnership. Global Water Partnership (GWP), Stockholm, Sweden

Breene K (2016) Food security and why it matters. Formative content. World Economic Forum. Available online at: https://www.weforum.org/agenda/2016/01/food-security-and-why-it-matters/. Accessed 22 Mar 2020

Brown C, Meeks R, Ghile Y et al (2013) Is water security necessary? An empirical analysis of the effects of climate hazards on national-level economic growth. Philos Trans R Soc A 371:20120416. https://doi.org/10.1098/rsta.2012.0416

C2ES (Center for Climate and Energy Solutions) (2011) Climate change 101: understanding and responding to global climate change: adaptation. Available online and accessed on 15.08.2020 at 03:20 at: https://www.c2es.org/site/assets/uploads/2017/10/climate101-fullbook.pdf and https://www.c2es.org/document/climate-change-101-understanding-and-responding-to-global-climate-change/

CAC (Codex Alimentarius Committee of the FAO/WHO) (2003) Codex Alimentarius, basic text on food hygiene, 3rd edn. FAO, Rome

Center for Development Research (ZEF), Forum for Agricultural Research in Africa (FARA), Institute of Agricultural Research for Development (IRAD) (2017) Country dossier: innovation for sustainable agricultural growth in Cameroon. Program of accompanying research for agricultural innovation. Center for Development Research, Forum for Agricultural Research in Africa and Institut de Recherche Agricole pour le Development, Bonn/Accra/Yaounde

Church J, Gregory J, Huybrechts P, Kuhn M, Lambeck K, Nhuan M, Qin D, Woodworth P (2001) Changes in sea level. In: Houghton J, Ding Y, Griggs D, Noguer M, van der Linden P, Xiaosu D (eds) Climate change 2001. The scientific basis. Cambridge University Press, Cambridge, pp 639–693

Convention on Biological Diversity (CBD) (2009) Connecting biodiversity and climate change mitigation and adaptation: report of the second ad hoc technical expert group on biodiversity and climate change. Technical series no. 41, Montreal, 126 p

Convention on Biological Diversity (CBD) (2018) Decision adopted by the conference of the parties to the convention on biological diversity: 14/5 biodiversity and climate change. CBD/COP/DEC/14/5

Dasgupta S, Laplante B, Meisner C et al (2009) The impact of sea level rise on developing countries: a comparative analysis. Clim Chang 93:379–388. https://doi.org/10.1007/s10584-008-9499-5

Direction de la Statistique et de la Comptabilité Nationale (DSCN) (2002) Conditions de vie des Populations et Profil de Pauvreté au Cameroun en 2001: Premiers resultats. Direction de la Statistique et de la Comptabilité Nationale, Yaoundé

Epule ET, Bryant RC (2016) Small scale farmers' indigenous agricultural adaptation options in the face of declining or stagnant crop yields in the Fako and meme divisions of Cameroon. Agriculture 6(2):22. https://doi.org/10.3390/agriculture6020022. Academic Editors: Annelie Holzkämper and Sibylle Stöckli (Special Issue Options for Agricultural Adaptation to Climate Change)

FAO (2010) The state of food insecurity in the world. Food and Agriculture Organization of the United Nations, Rome

Fonteh M, Esteves LS, Gehrels WR (2009) Mapping and valuation of ecosystems and economic activities along the coast of Cameroon: implications of future sea level rise. Int Approaches Coast Res Theory Pract Coastline Rep 13:47–63

Gabche CE, Smith VS (2002) Water, salts and nutrient budgets of two estuaries in the coastal zone of Cameroon. West Afr J Appl Ecol 3:69–89

Gommes R, du Guerny J (1998) Potential impacts of sea-level rise on populations and agriculture. Climate change in Food and Agriculture Organization of the United Nations. Available online at: http://www.fao.org/nr/climpag/pub/EIre0045_en.asp. Accessed 16 Mar 2020

Gopalakrishnan T, Hasan MK, Haque ATMS, Jayasinghe SL, Kumar L (2019) Sustainability of coastal agriculture under climate change. Sustainability 11:7200. https://doi.org/10.3390/su11247200

Ibe AC, Awosika LF (1991) Sea level rise impact on African coastal zones. In: Omide SH, Juma C (eds) A change in the weather: African perspectives on climate change. African Centre for Technology Studies, Nairobi, pp 105–112

Intergovernmental Panel on Climate Change (IPCC) (2001) Climate change 2001: the physical science basis. In: Houghton JT, Ding Y, Griggs DJ, Noguer M, van der Linden PJ, Xiaosu D (eds) Working group 1 contribution to the third assessment report of the Intergovernmental Panel on Climate Change (IPCC). Global Water Partnership (GWP), Stockholm, Sweden

Intergovernmental Panel on Climate Change (IPCC) (2007a) Climate change 2007 – mitigation of climate change contribution of working group iii to the fourth assessment report of the IPCC (978 0521 88011-4 Hardback; 978 0521 70598-1 Paperback)

Intergovernmental Panel on Climate Change (IPCC) (2007b) Climate change 2007 – the physical science basis contribution of working group I to the fourth assessment report of the IPCC (ISBN 978 0521 88009-1 Hardback; 978 0521 70596-7 Paperback)

Intergovernmental Panel on Climate Change (IPCC) (2007c) Climate change 2007 – impacts, adaptation and vulnerability contribution of working group II to the fourth assessment report of the IPCC (978 0521 88010-7 hardback; 978 0521 70597-4 paperback)

Intergovernmental Panel on Climate Change (IPCC) (2007d) Climate change 2007: synthesis report. In: Core Writing Team, Pachauri RK, Reisinger A (eds) Contribution of working groups I, II and III to the fourth assessment report of the intergovernmental panel on climate change. Global Water Partnership (GWP), Stockholm, Sweden

Intergovernmental Panel on Climate Change (IPCC) (2019) Special report on the ocean and cryosphere in a changing climate. In: Pörtner H-O, Roberts DC, Masson-Delmotte V, Zhai P, Tignor M, Poloczanska E, Mintenbeck K, Nicolai M, Okem A, Petzold J, Rama B, Weyer N (eds). In press. https://www.ipcc.ch/srocc/cite-report/

Jallow BP, Toure S, Barrow MMK, Mathieu AA (1999) Coastal zone of the Gambia and the Abidjan region in Cote d'Ivoire: sea level rise vulnerability, response strategies, and adaptation options. Clim Res 12(2):129–136. https://doi.org/10.3354/cr012129

Lambi CM, Kometa SS (2014) Climate change in Cameroon and its impacts on agriculture. In: Albrecht E, Schmidt M, Mißler-Behr M, Spyra S (eds) Implementing adaptation strategies by legal, economic and planning instruments on climate change. Environmental protection in the European Union, vol 4. Springer, Berlin/Heidelberg

Leal Filho W, F M, Nagy GJ et al (2018) Fostering coastal resilience to climate change vulnerability in Bangladesh, Brazil, Cameroon and Uruguay: a cross-country comparison. Mitig Adapt Strateg Glob Chang 23:579–602. https://doi.org/10.1007/s11027-017-9750-3

Living with Risk (2002) A global review of disaster reduction initiatives. International Strategy for Disaster Reduction (ISDR), Geneva. Accessed at: http://helid.digicollection.org/en/d/Js2653e/

Mimura N (2013) Sea-level rise caused by climate change and its implications for society. Proc Jpn Acad Ser B Phys Biol Sci 89:281–301

Molua EL (2006) Climatic trends in Cameroon: implications for agricultural management. Clim Res 30:255–262. https://doi.org/10.3354/cr030255

National Aeronautics and Space Administration (NASA) (2008) Ocean surface topography from space. Californian Institute of Technology: Jet Propulsion Laboratory, Pasadena, California

National Institute of Statistics (NIS) (2010) The population of Cameroon: reports of the presentation of the final results of the 3rd general census of population and habitats (RGPH): Central Bureau of Census and Population Studies. National Institute of Statistics, Ministry of the Economy and Finance, Yaounde

Nicholls RJ (2002) Analysis of global impacts of sea-level rise: a case study of flooding. Phys Chem Earth Parts A/B/C 27(32):1455–1466. https://doi.org/10.1016/S1474-7065(02)00090-6

Nicholls RJ, Tol RSJ (2006) Impacts and responses to sea-level rise: a global analysis of the SRES scenarios over the twenty-first century. Philos Trans R Soc A 364:1073–1095

Nicholls RJ, Wong PP, Burkett VR, Codignotto JO, Hay JE, McLean RF, Ragoonaden S, Woodroffe CD (2007) Coastal systems and low-lying areas. In: Parry ML, Canziani OF, Palutikof JP, van der Linden PJ, Hanson CE (eds) Climate change 2007: impacts, adaptation and vulnerability. Contribution of working group II to the fourth assessment report of the intergovernmental panel on climate change. Cambridge University Press, Cambridge, UK, pp 315–357

Nicholls RJ, Marinova N, Lowe JA, Brown S, Vellinga P, de Gusmão D, Hinkel J, Tol RSJ (2011) Sea-level rise and its possible impacts given a 'beyond 4°C world' in the twenty-first centuryPhil. Philos Trans R Soc A 369:161–181. https://doi.org/10.1098/rsta.2010.0291

Ogbuabor JE, Egwuchukwu EI (2017) The impact of climate change on the nigerian economy. Int J Energy Econ Policy, Econjournals 7(2):217–223.

Onguéné RE, Pemha E, Lyard F, Penhoat Y, Nkoue G, Duhaut T, Ebénézer N, Marsaleix P, Mbiake R, Jombe S, Allain D (2015) Overview of tide characteristics in Cameroon coastal areas using recent observations. Open J Mar Sci 05(1):81–98. https://doi.org/10.4236/ojms.2015.51008

Schiermeier Q (2018) Droughts, heatwaves and floods: how to tell when climate change is to blame. Nature 560:20–22

Stern N (2006) The economics of climate change. Cabinet office – HM treasury. Cambridge University Press, London

United Nations (UN) (2019) Climate change – United Nations. Available online at the DRIP POT SHOP: https://www.drippot.ca/blogs/climate-change-united/climate-change-united-nations. Accessed 16 Mar 2020

United Nations Environmental Program (UNEP) (2016) Options for ecosystem-based adaptation (EBA) in coastal environments: a guide for environmental managers and planners. UNEP, Nairobi

Worldometer (2020) Africa population forecast. Source: Worldometer (www.Worldometers.info). Available online and consulted on 07 Mar 2020 at: https://www.worldometers.info/world-population/africa-population/

2

Maize, Cassava and Sweet Potato Yield on Monthly Climate in Malawi

Floney P. Kawaye and Michael F. Hutchinson

Contents

Introduction .
Maize, Cassava, and Sweet Potato Growth
Methodology
   GROWEST Analysis
Results
   Trends in Production and Yield of Maize, Cassava, and Sweet Potato
   GROWEST Analyses

Conclusion
References

Abstract

Climate change and climate variability in Malawi have negatively affected the production of maize, a staple food crop. This has adversely affected food security. On the other hand, there have been increases in growing area, production, yield, consumption, and commercialization of both cassava and sweet potato. Factors behind these increases include the adaptive capacity of these crops in relation to climate change and variability, structural adjustment programs, population growth and urbanization, new farming technologies, and economic development. Cassava and sweet potato are seen to have the potential to contribute to food security and alleviate poverty among rural communities.

F. P. Kawaye (✉) · M. F. Hutchinson (✉)
Fenner School of Environment and Society, Australian National University,
Canberra, ACT, Australia
e-mail: Floney.Kawaye@anu.edu.au; michael.hutchinson@anu.edu.au

This study used a simple generic growth index model called GROWEST to model observed yields of maize, cassava, and sweet potato across Malawi between 2001 and 2012. The method can be viewed as a hybrid approach between complex process-based crop models and typical statistical models. For each food crop, the GROWEST model was able to provide a robust correlation between observed yields and spatially interpolated monthly climate. The model parameters, which included optimum growing temperatures and growing seasons, were well determined and agreed with known values. This indicated that these models could be used with reasonable confidence to project the impacts of climate change on crop yield. These projections could help assess the future of food security in Malawi under the changing climate and assist in planning for this future.

Keywords

Climate change · Food security · Maize · Cassava · Sweet potato · Crop yield modelling

Introduction

Process-based simulation models and statistical models are commonly used to assess the impact of climate variability and climate change on food crop yields (Lobell and Asseng 2017). The former typically have complex plant and environmental data requirements, while the latter can have large uncertainties in fitted parameters that make application to climate change assessment difficult (Schlenker and Lobell 2010; Ray et al. 2019). This chapter examines a hybrid approach that statistically calibrates a simple generic plant growth model, called GROWEST, using spatially distributed yield and monthly climate data. The model parameters are robustly determined and hence able to provide baseline models suitable for assessing the potential yields of maize, cassava, and sweet potato in relation to projected climate change. The GROWEST plant growth index model was originally developed by Fitzpatrick and Nix (1970) and Nix (1981) and has been implemented by Hutchinson et al. (2002). It has been used to develop a global agroclimatic classification that has identified agroclimatic classes for Australia (Hutchinson et al. 2005) that have been used to support a wide variety of ecosystem assessments.

The model is applied to annual yield data for the eight Agricultural Development Divisions (ADDs) across Malawi and corresponding spatially distributed monthly climate data. Spatial climate data were available from 1981 to 2012. The maize analyses were restricted to the years 2006–2012 after the introduction of the Farm Input Subsidy Program (FISP) in 2005/06 which had a significant impact on maize yields. Reliable annual yield data were available for cassava for the period 2001–2011 and for sweet potato for the period 2006–2012. The yield model is based on regressing the logarithms of the observed crop yields on average weekly growth indices of the GROWEST model, with the average taken over the respective growing season for each crop. Three key parameters of the GROWEST model, the optimum growth temperature and the starting and finishing weeks of the growth season, are tuned to the observed yield data for each crop. The regressions take into account variations in site and management conditions across different ADDs and are extended to take into account systematic increases in yields over time due to plant breeding programs, improvements in growing practice, and carbon dioxide fertilization. Robustly calibrated models are obtained for local, composite, and hybrid maize varieties and for cassava and sweet potato.

Maize, Cassava, and Sweet Potato Growth

As noted by Kawaye and Hutchinson (2018), maize is a staple food crop in Malawi that is grown under both irrigation and rainfed farming systems. However, rainfed farming dominates as it covers about 99% of Malawi's agriculture, mainly on smallholder farms. Kawaye and Hutchinson (2018) further noted that rainfed maize is normally planted between November and December (the start of the rainy season). It grows rapidly during the high rainfall months of January and February and matures by early April. It is harvested in late April or early May. Maize yields in Malawi range from less than 1000 kg to over 4000 kg per hectare, depending on various factors including climate, location, seed variety, fertilizer use, labor, and policy-related factors including access to credit, input and output markets and extension services.

Cassava (*Manihot esculenta* Crantz) is a perennial woody shrub with an edible starchy tuberous root (Mathieu-Colas et al. 2009). Cassava grows under diverse ecological and agronomical conditions. It favors a warm moist climate with mean temperature of 24–30 °C (Nassar 2004; Mkumbira 2002). It can tolerate temperatures from 16 °C to 38 °C (Cock 1984). It does not favor excess soil moisture nor high salt concentrations nor pH above 8 (Nassar 2004; Mkumbira 2002). Cassava is normally planted early in the wet season, usually around mid-November. As a perennial crop, cassava has no definite lifetime or maturation period. After full development of the canopy, root growth slowly decreases and finally stops. This is the maturation point of cassava when maximum or near maximum yield is obtained. Cassava is harvested when the returns for production and utilization are maximized. Thus harvesting can be delayed to well after when the tubers have matured. The optimum time to harvest is 9–12 months (Mathias and Kabambe 2015) depending on various ecological factors such as rainfall, temperature, and soil fertility (Mathieu-Colas et al. 2009; Benesi 2005).

Sweet potato (*Ipomoea batatas* Lam) is an annual crop (Mathieu-Colas et al. 2009). It is widely grown in tropical, subtropical, and temperate areas between 40°N and 32°S. It grows best with air temperatures between 20 °C and 25 °C, and growth is restricted below 15 °C (Ramirez 1992). It can be cultivated across a wide variety of soil types and prefers lightly acid or neutral soils with a pH between 5.5 and 6.5 (Ramirez 1992). Sweet potato is commonly grown as an intercrop with maize and planted early in the calendar year (FAO 2005). It can also be planted toward the end of the wet season in late April and grown on residual soil moisture. Like cassava, the growth period of sweet potato depends on various ecological factors, but it generally takes 4–5 months to mature (Mathieu-Colas et al. 2009).

Methodology

This study modelled the dependence of observed maize, cassava, and sweet potato yields on spatially distributed monthly climate to provide a basis for assessing the impact of projected climate change. The models were constructed by performing multilinear regressions of the logarithms of the observed yields on accumulated outputs from the GROWEST growth index model. The GROWEST model generates a generic, process-based, growth index that depends on weekly or monthly climate with a minimal number of parameters. The growth index varies between 0 (climate totally limiting for growth) and 1 (climate optimal for growth). It is normally assumed to be proportional to the rate of relative increase in plant biomass over time. The growth index is calculated as the product of three separate indices that incorporate the impact of temperature, solar radiation, and modelled soil moisture. The simplicity of the model, and its underlying process basis, makes it well suited to deriving robust calibrations of yield response to climate using yield data limited in quantity or quality. The modelling approach can be seen as a hybrid between complex crop simulation models and statistical analysis of growing season weather variables (Lobell et al. 2011). The critical parameters of the GROWEST model are able to be determined by maximizing the alignment of the GROWEST outputs with the observed crop yields using standard multilinear regression. These regression analyses are extended to take account of observed systematic increases in crop yields over time due to crop breeding programs, improvements in crop management, and possible carbon dioxide fertilization.

GROWEST Analysis

The models were calibrated on yield data for each ADD over the periods for which both climate and reliable yield data were available, namely, 2006–2012 for the three main varieties of maize, 2001–2011 for cassava, and 2006–2012 for sweet potato. The GROWEST model was applied to monthly climate across all eight ADDs using interpolated monthly climate values at 74 points that approximately equi-sampled the cropping areas across the eight ADDs of Malawi as shown in Fig. 1. The interpolated climate values were obtained using thin plate smoothing splines as provided by the ANUSPLIN Version 4.4 package (Hutchinson and Xu 2013). The output growth indices (GIs) were averaged across each ADD to match the average yields reported for each ADD. The GROWEST model runs on a weekly time step but can be applied to monthly climate data by interpolating monthly data to weekly

Fig. 1 Seventy-four sites
sampling crop growing areas
across the eight Agricultural
Development Divisions
(ADDs) of Malawi

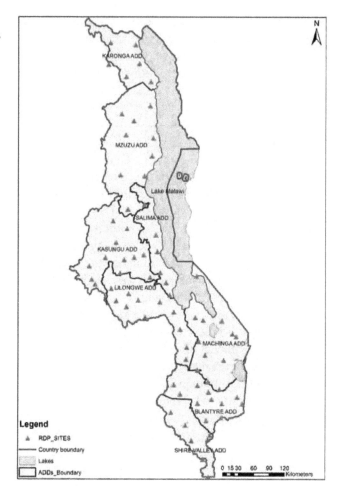

data "on the fly." It can provide weekly or monthly outputs. Weekly GROWEST
outputs were used to obtain finer-scale demarcations of the fitted growing seasons.

A robust regression model was used to determine critical GROWEST parameters
for all five crop varieties. Since the GROWEST model describes a relative growth rate,
it is natural to formulate the regression model in terms of the natural logarithm of the
observed yields. The log formulation has been used in other yield studies (Lobell et al.
2011) and appears to offer a natural separation between site and climatic effects on
observed crop yields. A similar approach, but not applying the log transformation, was
described by Kawaye and Hutchinson (2018). The regression model for the observed
yields Z_{ij} for year Y_i and ADD_j ($j = 1,..,8$) was defined by:

$$\log \left(Z_{ij} \right) = a_j + b \, G_{ij} + c Y_i + \varepsilon_{ij}$$

where G_{ij} denotes the accumulated growth index for year Y_i and ADD_j and ε_{ij}
denotes a zero-mean random error. The model parameters a_j and b were initially

fitted by least squares regression while setting c = 0. Three additional critical GROWEST parameters were optimized during this initial fitting of the model. These parameters were the optimum temperature of the temperature index, used in calculating the weekly growth index, and the first and last weeks of the fitted growing season, used to define the growing season period. The soil water balance parameters of the GROWEST model were set to default values with the soil water holding capacity set to 150 mm and the soil drying rate set to that of a clay loam soil. The soil water balance is an important component of the growth index, but it is not very sensitive to departures from these default parameter values.

For the three maize varieties, parameter c was fitted by refitting all parameters using least squares regression. For cassava and sweet potato, parameter c was fitted by least squares regression on year of the residuals of the data from the initial model. This removed instabilities when all parameters were fitted directly to the cassava and sweet potato data.

The model is robust with just ten parameters. It has a constant dependence of log (yield) on the climate-based accumulated growth index but has a site varying intercept to allow for different site and management conditions across different ADDs. The linear dependence on year via parameter c allows for underlying improvements in yield due to crop breeding programs, improvements in crop management, and possible carbon dioxide fertilization.

As in Kawaye and Hutchinson (2018), the GROWEST parameters were optimized by automating GROWEST runs and initial regressions using FORTRAN code and standard LINPACK numerical analysis software (Dongarra et al. 1979). Comprehensive analyses of the fitted models were computed using the standard regression package within Excel software. These analyses permitted the identification and removal of a small number of yield data outliers with large standardized residuals. These were associated with anomalies and accounting errors evident in the supporting yield data.

Once outliers were removed and GROWEST optimizations were complete, final comprehensive statistical analyses were computed, making due allowance for the three degrees of freedom associated with estimating the three GROWEST parameters. Standard errors of the fitted GROWEST parameters for the initial regressions were calculated from the diagonal elements of the inverse of the associated Hessian matrix. The Hessian matrix was estimated by calculating second-order finite differences of the residual sums of squares of the model with respect to the three fitted GROWEST parameters.

The final fitted models were applied to the associated growing areas for each ADD and each year. These outputs were aggregated across all ADDs to obtain modelled national yield and production for all crops analyzed. These values were compared with the tabulated national yield and production values to assess the performance of the models at the national scale.

With a view to assessing the potential impacts of climate change, the domains of the fitted temperature indices for each crop were compared with the distributions of the weekly temperatures that occurred across the ADDs over the fitted growing seasons.

Results

Trends in Production and Yield of Maize, Cassava, and Sweet Potato

Three main varieties of maize are cultivated in Malawi. These are (i) local (traditional) varieties, (ii) composite varieties, and (iii) hybrid varieties. There are major differences between the yield potentials of these varieties (Giertz et al. 2015; Pauw et al. 2010; Denning et al. 2009; JICAF 2008; Heisey and Smale 1995; Ngwira and Sibale 1986). Local varieties have the lowest yield. They are not subject to yield improvement programs, and harvested seed is recycled from year to year. Composite varieties have higher yields and are often more drought tolerant. They are subject to yield improvement programs but seed can be recycled. Hybrid varieties are the highest yielding and most expensive. They are subject to strictly controlled yield improvement programs, and harvested seed cannot be recycled.

Figure 2 compares growing areas and yields of these three maize varieties from 1984 to 2015. There has only been significant composite maize production since the late 1990s. There has been a steady increase in the area devoted to composite and hybrid varieties and a simultaneous reduction in local maize growing area. This shift has been encouraged by increasing climate stress, such as increasing temperatures, and poor access to farm inputs for local maize production. There is significant year-to-year variation in maize yields, with composite and hybrid yields particularly low from 2001 to 2005. This could be attributed in part to poor climate including low rainfall. The generally higher yields of composite and hybrid maize after 2005 coincide with the introduction of the Farm Input Subsidy Program (FISP). Kawaye and Hutchinson (2018) have presented evidence that FISP has made a significant improvement in composite and hybrid yields since 2006. The analysis of maize yields presented below is therefore restricted to the post FISP years.

Figure 3 shows that cassava and sweet potato production has been generally increasing. There was an abrupt increase in cassava yield in the year 2000 followed by a steady increase, while sweet potato yield has been steadily increasing since the mid-1990s. The abrupt increases in yields in earlier years suggest there have been major improvements in crop-growing practice, and perhaps recording practice, during the 1990s. The steady increase in yields since 2000 reflects increased policy and institutional support, such as the introduction of higher yielding varieties, and improved management practices to diversify the food security basket. The general increase in area under cultivation cassava and sweet potato since 2005 indicates that more farmers have been planting these crops on new land or on land withdrawn from or shared with maize. As noted above, this is due to an increasing reliance on cassava and sweet potato for food security, especially in maize deficit (drought) years.

There was a sharp drop in both yield and production for all crops (maize, cassava, and sweet potato) in the drought year of 2005. For the other years, major variations in production are largely explained by major variations in growing areas as shown in Figs. 2 and 3. On the other hand, minor year-to-year variations in yield are likely to be attributable to year-to-year climatic variations. The differing year-to-year

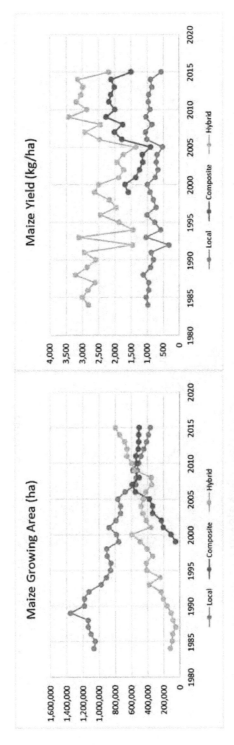

Fig. 2 Growing areas and total yield for local, composite, and hybrid maize from 1984 to 2015

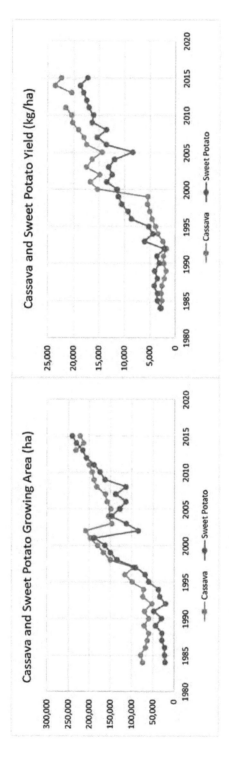

Fig. 3 Growing areas and total yield of cassava and sweet potato from 1984 to 2015

variations in yield and climate by ADD enable the calibration of the yield responses to monthly climate using the models described here.

GROWEST Analyses

For each crop the model was initially fitted to the available yield data values with the optimum GROWEST temperature set to a commonly accepted value for each crop. The automated model fitting code was used to adjust the growing season to minimize the standard error of the fitted model. A small number of large standardized residuals for each model were found to correspond to anomalous growing areas or yields for particular RDPs making up the ADD yield data. These large outliers were removed from the analysis, and the three key GROWEST parameters, the optimum growing temperature and the beginning and end of the growing season, were finally optimized by minimizing the residual sum of squares of the fitted model. There were two outliers for each of the three maize analyses, none for the cassava analysis and four for the sweet potato analysis. The values of the three key GROWEST parameters were found to be quite critical to the overall performance of the model. This is in keeping with the finding of Wang et al. (2017) who similarly found that the shape and location of the temperature response function in process-based crop models are critical to their performance. The fitted GROWEST parameters are listed in Table 1.

The fitted starting weeks for the three maize varieties were remarkably similar. Allowing for a period of around 2 weeks from sowing to emergence when the maize plants begin to interact with the atmosphere, the fitted starting weeks corresponded to planting in early December when the wet season is underway. Composite and hybrid maize had fitted growing seasons lasting 13 and 12 weeks, while local maize had a somewhat longer fitted growing season of 15 weeks. These values are all in reasonable agreement with standard management practice. There is systematic variation in the fitted optimum growing temperatures. The fitted temperature for local maize is consistent with documented optimum temperatures of around 26 °C

Table 1 Optimized GROWEST parameters and mean yield for each crop. Standard errors for each parameter value are provided in parentheses. Weeks are numbered sequentially 1–52 for each year

Crop	Number of data points	Optimum growing temperature (°C)	First week of growing season	Last week of growing season	Mean crop yield (kg/ha)
Local maize	54	24.5 (0.9)	52 (1.0)	15 (1.7)	928
Composite maize	54	28.9 (0.4)	52 (0.8)	13 (1.3)	1,981
Hybrid maize	54	30.2 (0.6)	51 (1.4)	12 (1.5)	2,883
Cassava	88	27.3 (0.6)	47 (2.0)	26 (2.9)	17,984
Sweet potato	52	23.4 (0.7)	7 (2.9)	22 (2.0)	15,920

for maize root growth and grain filling (Sánchez et al. 2014). The fitted temperature for composite maize is consistent with documented optimum temperatures of around 28 °C to 30 °C for maize growth from sowing to anthesis. The slightly higher fitted temperature for hybrid maize is consistent with documented optimum temperatures of around 31 °C for whole plant maize growth. The fitted parameter values indicate that hybrid and composite maize are better adapted to higher temperatures and have shorter growing seasons than the traditional local maize varieties. These are both accepted aims of maize breeding programs.

The fitted growing season for cassava corresponded to planting cassava in mid-November and effective growth terminating by around the end of June. This agrees with the usual growing practice for cassava reported above, with planting time somewhat variable from year to year depending on the arrival of rain and harvesting time variable according to a range of conditions. The latter is consistent with the larger standard error for the finishing week. The fitted optimum growing temperature for cassava is in good agreement with accepted values (Nassar 2004; Mkumbira 2002; Cock 1984).

The fitted growing season for sweet potato corresponded to planting sweet potato in early February and effective growth terminating by around the beginning of June. This is consistent with sweet potato being mainly grown as an intercrop with maize and planted after the maize crop is in place. The fitted optimum growing temperature for sweet potato is also in good agreement with accepted values (Ramirez 1992).

Statistics and key model parameter estimates for each crop are provided in Table 2. All of the model fits were highly statistically significant well beyond the 0.001% level. The performance of the model, with a single coefficient of the accumulated growth index and a different model intercept for each ADD, is remarkably consistent across all crops. Allowing the coefficient of the accumulated growth index to vary from ADD to ADD gave unstable behavior and did not improve the standard error of any model. This confirmed that the relative dependence of crop yield on climate via the accumulated growth index was effectively constant across all sites, justifying the use of a single parameter b across all ADDs.

Table 2 Critical parameter estimates and the percent of yearly variance accounted for the fitted models for all five crops. Standard errors of the fitted parameter values are provided in parentheses

Crop	Number of data points	Model standard error	Parameter b	Parameter c	Percent yearly variance accounted for (%)
Local maize	54	0.141	2.79 (0.55)	0.000 (0.009)	38
Composite maize	54	0.141	3.33 (0.71)	0.013 (0.010)	36
Hybrid maize	54	0.150	2.61 (0.69)	0.032 (0.011)	34
Cassava	88	0.132	2.16 (0.34)	0.025 (0.004)	54
Sweet potato	52	0.070	0.85 (0.17)	0.016 (0.005)	47

The standard errors of the fitted models are generally no more than 15%, and the fitted values of parameter b are mostly between 2 and 3 with relatively small standard errors of around 20%. The percent of yearly variation in crop yields explained by the model, after removing variation between sites, ranged from 34% for hybrid maize to 54% for cassava. These values are consistent with the finding of Ray et al. (2015) that climate variation explains around a third or more of crop yield variability.

The smaller value of parameter b for sweet potato suggests that the model has been less successful in calibrating the full impact of climate on tuber growth. This may have been contributed to by the relatively short data record available for sweet potato and the larger number of apparent accounting errors in the yield data. The variation in planting dates between the traditional early planting date in February and the less common late planting date at the end of the wet season may have also contributed to the less strong fitted dependence on climate via parameter b. On the other hand, the small model standard error suggests that sweet potato yields may be more stable in relation to climatic variability than cassava. Analysis of yield data over a larger number of years would help to resolve this question.

The underlying rates of increase in crop yields are well determined for all crops. The fitted rate of zero for local maize is consistent with no breeding program in place for local maize. The marginally statistically significant rate of increase for composite maize of 1.3% per year is consistent with the modest breeding program in place for composite maize, and the statistically significant rate for hybrid maize of 3.2% per year is consistent with the strong breeding program in place for hybrid maize. The fitted rates of increase of around 2% per year for sweet potato and cassava are consistent with breeding programs being in place for both crops.

Plots of the log (yield) data values versus modelled values are shown in Fig. 4. Individual plots (not shown) of the fitted model, as a function of accumulated GI for each ADD, show considerable scatter of the observed yield data about the fitted model, but the constant slope b of the fitted line across all ADDs is estimated with reasonable precision, as described above. Likely contributors to the scatter about the fitted model include changes in management practice from year to year, including variations in planting dates, inaccuracies in recording crop yields, and possible misalignments between the locations of the sites sampling climate across the ADDs and the main locations of crop growth. The monthly time scale of the supporting climate data is likely to have made only a small contribution to the scatter about the fitted model given that the largest departures from the fitted model are in every case attributable to clear accounting errors in particular resource development districts (RDDs) within each ADD rather than systematic climate-related anomalies across all RDDs in any ADD.

The smallest observed and modelled values in the plot for cassava are for the drought year 2005 in Shire Valley ADD. The drought was severe in the other southern Machinga and Blantyre ADDs but particularly severe for the Shire Valley ADD (FAO 2005). The plots in Fig. 4 show that the model is able to recognize most drought conditions with reasonable accuracy but somewhat overestimates cassava yield during particularly severe droughts.

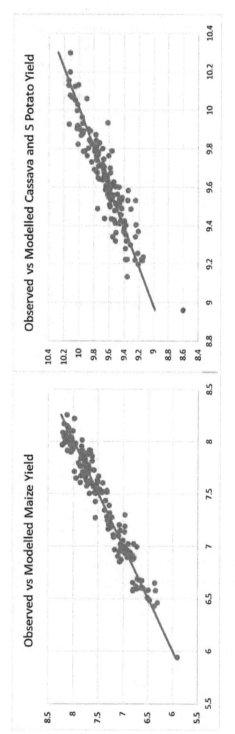

Fig. 4 Log (ADD yield) versus log (modelled ADD yield) for maize, cassava, and sweet potato

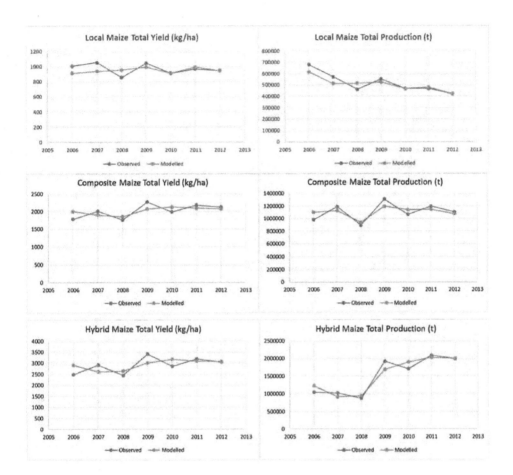

Fig. 5 Modelled and actual total yield and production for all three maize varieties

The models were finally assessed by their ability to explain the total yield and production across Malawi. The observed and modelled yield data were aggregated across all ADDs and plotted in Figs. 5 and 6. The spatially aggregated models provide accurate explanations of the observed values with average percentage differences from the actual annual values around 7% for the three maize varieties and less than 5% for cassava and sweet potato. The larger departures in these plots generally correspond with known accounting errors in the supporting yield data.

The fitted temperature index curves and the corresponding relative histogram of weekly temperatures observed over the fitted growing season for the 74 sites representing the eight ADDs across Malawi are plotted for local maize, hybrid maize, cassava, and sweet potato in Fig. 7. The plots show that the apparent temperature constraints on local maize and sweet potato are well matched to the observed weekly temperatures across Malawi, while the optimum temperatures of hybrid maize and cassava are somewhat larger than the mode of the observed weekly average temperatures. This suggests that projected future increases in temperature of

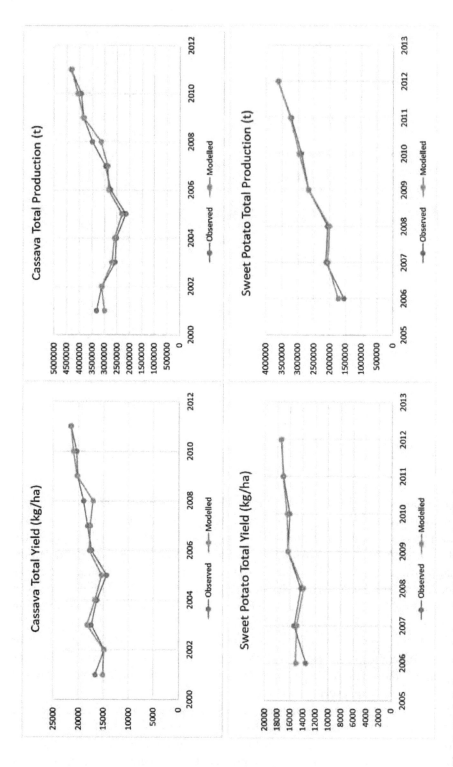

Fig. 6 Modelled and actual total yield and production for cassava and sweet potato

around 2 °C would have minimal impact on local maize yield, perhaps slightly reduce sweet potato yield and moderately enhance yields of higher temperature adapted hybrid maize and cassava. Possible changes in soil moisture regimes also need to be taken into account to obtain a more complete estimate of the likely impact of projected future climate. Soil moisture status is particularly important in the grain filling stage of maize (Li et al. 2018).

Discussion

The fitted regressions on accumulated GI have provided reasonably accurate models of observed maize, cassava, and sweet potato yields for each ADD and are more accurate when aggregated to the national level. The model formulation is robust and able to fit well-determined trends on accumulated growth index, despite the uncertainty associated with the supporting data, including some imprecision in the location of actual crop growing areas and year-to-year variations in planting times and growing practice. Schlenker and Lobell (2010) have noted the particular difficulties in modelling cassava that this model has appeared to overcome.

The fitted growing seasons agree well with known practice. The fitted optimum temperatures also agree with generally accepted values for all five crops, with hybrid maize better adapted to higher temperatures than traditional local maize varieties. This close agreement with known values provides strong support for the adequacy of the fitted models in calibrating the climate dependencies of maize, cassava, and sweet potato yields. The formulation of the regression model has permitted an effective separation between site-specific effects (such as soil fertility and particular crop management practices) and climatic effects on relative plant growth. The site-specific effects are accounted for by a separate intercept for each ADD in the regression model, while the relative climatic effects appear to operate independently of different site conditions and can be effectively calibrated by a single factor across all ADDs. Allowing this factor to vary across the ADDs did not improve the fit of the model for any crop. The model formulation is similar to that employed by Lobell et al. (2011) but uses a specifically tuned nonlinear plant growth index instead of various growing season weather variables. The effectiveness of this modelling approach reflects the finding of Wang et al. (2017) that the form of the temperature response function is quite critical in the accuracy of crop simulation models. The functional form of the temperature indexes plotted in Fig. 7 is similar to the preferred functional forms described by Wang et al. (2017).

The single parameter for the climatic effects was an important factor in the robustness of the regression growth models. On the other hand, allowing a separate site-specific intercept for each ADD was an important factor in incorporating different conditions modifying yields across the different ADDs. The resulting robust statistical model could reliably detect data outliers, as confirmed by inspection of the supporting data for the contributing RDPs. The robustness of the spatial analyses of the supporting monthly climate data has also contributed to the robustness of the fitted growth models. The net result has been well-determined

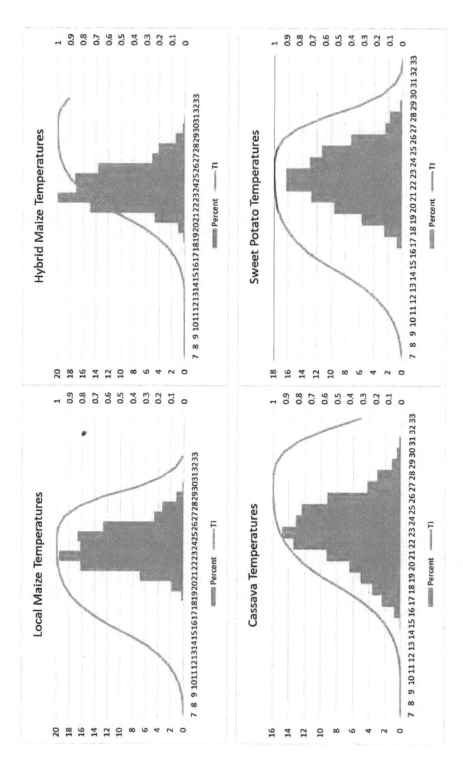

Fig. 7 Fitted temperature index curves (line) and observed monthly average temperatures (relative histogram) over the fitted growing season for local maize, hybrid maize, cassava, and sweet potato

coefficients calibrating the dependence of the three maize varieties and cassava and sweet potato yields on monthly climate via appropriately accumulated GI.

The models have simultaneously calibrated significant underlying increasing trends in yield over time that can be ascribed to improvements in plant breeding, crop management, and carbon dioxide fertilization. The fitted underlying trends of around 2% increase in yield per year for cassava and sweet potato may have been augmented by carbon dioxide fertilization. For such tubers there is around a 15% increase in tuber yield per 100 parts per million increase in atmospheric carbon dioxide concentration (Kimball 1983; Miglietta et al. 1998). In view of the prevailing rate of an increase in carbon dioxide concentration of two parts per million per year, this would give rise to an increase in tuber yield of around 0.3% per year over the analysis period. This is similar to the standard error of the fitted annual percentage increases in crop yield, making it difficult to discriminate from other increases in crop yield. It is not clear whether the fitted underlying percentage increases in crop yields will be maintained indefinitely. The impacts of improved production and reporting methods are likely to plateau in the future. However, ongoing improvements due to plant breeding and carbon dioxide fertilization are likely to continue, and the fitted trends of around 2% per year are remarkably consistent with the documented increase in world average cereal yields over the 50 years from 1961 to 2009 (Prohens 2011). It is unlikely that conventional crop breeding methods are able to maintain this rate of progress into the future, but Prohens (2011) argues that recent progress in molecular biology and genetic engineering offers great promise to further increase crop yields. Thus the fitted models should be able to be used, with appropriate qualifications, in assessing the impact of projected climate change.

Conclusion

This study analyzed the impact of monthly climate on the observed yields of maize, cassava, and sweet potato across the eight ADD crop production regions of Malawi via a robust yield regression model that can be viewed as a hybrid approach between complex process-based models and statistical modelling using selected weather variables. It offers progress toward the eventual dissolution of the differences between these approaches as suggested by Lobell and Asseng (2017). A particular strength of the GROWEST model used here is its incorporation of a process-based temperature response function that could be readily tuned to maximize model performance. This reflects the finding of Wang et al. (2017) that appropriate parameterization of the temperature response function is critical to crop model performance. The tuning of the period of the effective growing season was similarly critical. An additional important contribution to the accuracy and robustness of the regressed GROWEST models is their effective separation of site-specific and climatic effects. This aspect is shared by the statistical modelling approach described by Lobell et al. (2011). Finally, the calibration of the model in terms of monthly, instead of daily, climate data offers robustness in delivering spatially distributed climate data from limited point sources and in generating projected future climate data. Projecting daily precipitation data in particular is problematic with many approaches simply adjusting positive daily rainfall amounts but leaving daily rainfall occurrence structure unchanged. Such changes in daily precipitation structure can be subsumed within simple changes in monthly precipitation totals, although at the expense of losing some precision in the timing of precipitation within the month.

The main limitation of the modelling approach described here, and many others, is an assumption that management practice does not change significantly from year to year. This can be violated in periods of extreme drought or flooding when planting dates can be significantly delayed or disrupted. This assumption could also be violated in future scenarios when there could be a systematic shift in planting times in response to systematic changes in seasonal climate. Changes in planting density due to changes in intercropping practice are also possible. There is also an assumption that a fixed temperature response function applies over the entire growing season. The differing temperature optima over different stages of maize growth cited by Sánchez et al. (2014) suggest that the model could be usefully elaborated to reflect this, although at the expense of fitting additional model parameters. Despite these limitations, the GROWEST plant growth index model applied to spatially distributed monthly climate data has provided robust correlations between modelled and actual yields for all five crops examined. These correlations have yielded process-based parameter values that agree with known values, and the dependence on accumulated growth index has been fitted with enough precision for the models to be able to be used with reasonable confidence in projecting the impacts of climate change on future yields. The comparisons of the fitted temperature index curves with observed monthly average temperatures in Fig. 7 show that projected increases in temperature are likely to have minimal impact on local maize and sweet potato yield while yields of high temperature adapted hybrid maize and cassava are likely to be enhanced. Such projections need to be coordinated with projected changes in soil moisture levels.

References

Benesi IR (2005) Characterization of Malawian cassava germplasm for diversity, starch extraction and its native and modified properties. PhD thesis, Department of Plant Science, University of the Free State, Bloemfontein

Cock JH (1984) Cassava. In: Goldsworthy PR, Fisher NM (eds) The physiology of tropical field crops. Wiley, Chichester, pp 529–550

Denning G, Kabembe P, Sanchez P, Malik A, Flor R, Harawa R, Nkhoma P, Zamba C, Banda C, Magombo C, Keating M, Wangila J, Sachs J (2009) Input subsidies to improve smallholder maize productivity in Malawi: toward an African green revolution. PLoS Biol 17(1):e23

Dongarra JJ, Moler CB, Bunch JR, Stewart GW (1979) LINPACK user's guide. SIAM Publications, Philadelphia

Fitzpatrick EA, Nix HA (1970) The climatic factor in Australian grassland ecology. In: Moore RM (ed) Australian grasslands. ANU Press, Canberra

Food and Agriculture Organization (FAO) (2005) FAO/WFP Crop and Food Supply Assessment Mission to Malawi, June 2005. http://www.fao.org/3/J5509e/J5509e00.htm

Giertz A, Caballero J, Galperin D, Makoka D, Olson J, German G (2015) Malawi: agricultural sector risk assessment. World Bank Group report 99941-MW. World Bank, Washington, DC

Heisey PW, Smale M (1995) Maize technology in Malawi: a green revolution in the making? CIMMYT research report no. 4. CIMMYT, Mexico

Hutchinson MF, Xu T (2013) ANUSPLIN version 4.4 user guide. Fenner School of Environment and Society, The Australian National University. https://fennerschool.anu.edu.au/research/products/anusplin

Hutchinson MF, Nix HA, McMahon JP McTaggart C (2002) GROWEST version 2.0. Centre for Resource and Environmental Studies. Australian National University, Canberra. https://fennerschool.anu.edu.au/research/products/growest

Hutchinson MF, McIntyre S, Hobbs RJ, Stein JL, Garnett S, Kinloch J (2005) Integrating a global agro-climatic classification with bioregional boundaries. Glob Ecol Biogeogr 14:197–212

Japan Association for International Collaboration of Agriculture and Forestry (JICAF) (2008) The Maize in Zambia and Malawi. Accessed 10 Mar 2017. http://www.apip-apec.com/ja/good-practices/files/The_Maize_in_Zambia_and_Malawi.pdf

Kawaye FP, Hutchinson MF (2018) Are increases in maize production in Malawi due to favourable climate or the farm input subsidy program? In: Alves F, Filho WL, Azeiteiro U (eds) Theory and practice of climate adaptation. Springer, Cham, pp 375–390

Kimball BA (1983) Carbon dioxide and agricultural yield: an assemblage and analysis of 430 prior observations. Agron J 75:779–788

Li Y, Tao H, Zhang B, Huang S, Wang P (2018) Timing of water deficit limits maize kernel setting in association with changes in the source-flow-sink relationship. Front Plant Sci 9:1326

Lobell DB, Asseng S (2017) Comparing estimates of climate change impacts from process-based and statistical crop models. Environ Res Lett 12:015001

Lobell DB, Bänziger M, Magorokosho C, Vivek B (2011) Nonlinear heat effects on African maize as evidenced by historical yield trials. Nat Clim Chang 1:42–45

Mathias L, Kabambe VH (2015) Potential to increase cassava yields through cattle manure and fertilizer application: results from Bunda College, Central Malawi. African J Plant Sci 9(5):228–234

Mathieu-Colas L, Bahers G, Makina LM (2009) CROP MANAGEMENT: roots and tubers trainer's guideline. Fiche AGRO – Cultures, Inter Aide Agro-Phalombe, Malawi

Miglietta F, Magliulo V, Bindi M, Cerio L, Vaccari FP, Loduca V, Peressotti A (1998) Free air CO_2 enrichment of potato (*Solanum tuberosum* L.): development, growth and yield. Glob Chang Biol 4:163–172

Mkumbira J (2002) Cassava development for small scale farmers: approaches to breeding in Malawi. Doctoral thesis. Swedish University of Agricultural Sciences, Uppsala, Sweden. Acta Universitatis Agriculturae Sueciae. Agraria 365. Tryck:SLU Service/Repro, Uppsala

Nassar NMA (2004) Cassava: some ecological and physiological aspects related to plant breeding. Gene Conserv 3(13):229–245

Ngwira LDM, Sibale EM (1986) Maize research and production in Malawi. In: Gelaw B (ed) To feed ourselves: proceedings of the first eastern, central, and southern African regional maize workshop. Mexico, CIMMYT

Nix HA (1981) Simplified simulation models based on specified minimum data sets: the CROPEVAL concept. In: Berg A (ed) Application of remote sensing to agricultural production forecasting. Commission of the European Communities, Rotterdam, pp 151–169

Pauw P, Thurlow J, van Seventer D (2010) Droughts and floods in Malawi. IFPRI discussion paper 00962. International Food Policy Research Institute, Washington, DC

Prohens J (2011) Plant breeding: a success story to be continued thanks to the advances in genomics. Front Plant Sci 2:Article 51

Ramirez P (1992) Cultivation harvesting and storage of sweet potato products. In: Machin D, Nyvold S (eds) Roots, tubers, plantains and bananas in animal feeding. Proceedings of the FAO Expert Consultation held in CIAT, Cali, Colombia 21–25 January 1991; FAO animal production and health paper – 95

Ray DK, Gerber JS, MacDonald GK, West PC (2015) Climate variation explains a third of global crop yield variability. Nat Commun 6:5989

Ray DK et al (2019) Climate change has likely already affected global food production. PLoS One 14(5):e0217148

Sánchez B, Rasmussen A, Porter JR (2014) Temperatures and the growth and development of maize and rice: a review. Glob Chang Biol 20:408–417

Schlenker W, Lobell DB (2010) Robust negative impacts of climate change on African agriculture. Environ Res Lett 5(1):014010

Wang E et al (2017) The uncertainty of crop yield projections is reduced by improved temperature response functions. Nat Plants 3:17102

3

Risks of Indoor Overheating in Low-Cost Dwellings on the South African Lowveld

Newton R. Matandirotya, Dirk P. Cilliers, Roelof P. Burger, Christian Pauw and Stuart J. Piketh

Contents

Introduction ....
Materials and Methods
    Description of the Study Site and Sampled Dwellings
    Description of Indoor Temperature Monitoring Sensors: (*Thermochron iButton*
    Indoor Air Temperature Measurements (T_{ai}
    Ambient Air Temperature Measurements (T_{air}
    Data Analysis
Results

    Study Limitations
Conclusion
References

Abstract

The South African Lowveld is a region of land that lies between 150 and 2000 m above sea level. In summer the region is characterized by the maximum mean daily ambient temperature of 32 °C. The purpose of the study was to characterize

N. R. Matandirotya (✉) · D. P. Cilliers · R. P. Burger · S. J. Piketh
Unit for Environmental Sciences and Management, North-West University,
Potchefstroom, South Africa
e-mail: runyamore@gmail.com; Dirk.Cilliers@nwu.ac.za; roelof.burger@nwu.ac.za; Stuart.
Piketh@nwu.ac.za

C. Pauw
NOVA Institute, Pretoria, South Africa
e-mail: christiaan.pauw@nova.org.za

indoor thermal environments in low-cost residential dwellings during summer seasons as climate is changing. Indoor and ambient air temperature measurements were performed at a 30-min temporal resolution using Thermochron *i*Buttons in the settlement of Agincourt. 58 free running low-cost residential dwellings were sampled over the summer seasons of 2016 and 2017. Complementary ambient air temperature data were sourced from the South African Weather Service (SAWS). Data were transformed into hourly means for further analysis. It was found that hourly maximum mean indoor temperatures ranged between 27 °C (daytime) and 23 °C (nighttime) for both living rooms and bedrooms in summer 2016 while in 2017, maximum mean indoor temperatures ranged between 29 °C (daytime) and 26 °C (nighttime) in living rooms and bedrooms. Pearson correlations showed a positive association between indoor and ambient temperatures ranging between $r = 0.40$ (daytime) and $r = 0.90$ (nighttime). The association is weak to moderate during daytime because occupants apply other ventilation practices that reduce the relationship between indoor and ambient temperatures. The close association between nighttime ambient and indoor temperature can also be attributed to the effect of urban heat island as nighttime ambient temperature remain elevated; thus, influencing indoor temperatures also remain high. These findings highlight the potential threat posed by a rise in temperatures for low-cost residential dwellings occupants due to climate change. Furthermore, the high level of sensitiveness of dwellings to ambient temperature changes also indicates housing envelopes that have poor thermal resistance to withstand the Lowveld region's harsh extreme heat conditions, especially during summer. The study findings suggest that a potential risk of indoor overheating exists in low-cost dwellings on the South African Lowveld as the frequency and intensity of heat waves rise. There is therefore a need to develop immediate housing adaptation interventions that mitigate against the projected ambient temperature rise for example through thermal insulation retrofits on the existing housing stock and passive housing designs for new housing stock.

Keywords

Low-cost dwellings · Indoor temperatures · Climate change · Indoor human thermal comfort · Thermal discomfort · Residential dwellings · Extreme weather · Adaptation · Overheating

Introduction

Climate change has triggered a rise in external summer air temperature posing a threat to human indoor thermal comfort and health (Mavrogianni et al. 2010) with cities facing a heightened risk of extreme heat from climate change (Araos et al. 2016). On a regional scale temperature rise in sub-Saharan Africa is expected to be higher than global mean temperatures (Webber et al. 2018; Hoegh-Guldberg et al. 2018) as Southern Africa is expected to rise at 2 °C compared to a global mean of

1.5 °C (Hoegh-Guldberg et al. 2018). Incidences of extreme weather events around the world are increasing as a consequence of climate change with projections showing an increase in the number of warm days/nights putting low-icome earners at risk of physical health threats (Hoegh-Guldberg et al. 2018). The most severe effects of global warming will be reflected through an increase in the frequency and intensity of extreme events such as heatwaves (IPCC 2007; White-Newsome et al. 2012; Dosio 2016). Low-income earners who live in substandard housing are likely to feel the full wrath of temperature rises as they cannot adapt quickly to climatic change.

According to the Chartered Institution of Building Services Engineers (CIBSE Guide A), sleep impairment can be experienced at temperatures above 24 °C (Mavrogianni et al. 2010). Low-cost residential dwellings that have poor thermal insulation can overheat if exposed to extreme heat. High indoor temperatures also drive energy consumption via the occupants' demand for cooling (Kavgic et al. 2012). Low-cost residential dwellings in South Africa are poorly insulated exposing them to extreme heat effects (Chersich et al. 2018). During hot spellslow-cost structures may be 4–5 °C warmer than outdoor temperatures (Chersich et al. 2018). Indoor over-heating is a function of dwelling thermal insulation levels as well as ventilation practices of occupants (Mavrogianni et al. 2010). Overheating occurs when the indoor operative temperature is over 3 °C the thermal comfort temperature (Mavrogianni et al. 2010) which the WHO pegged at 24 °C for indoor environments.

Similar studies on summer indoor temperature monitoring include studies by Summerfield et al. (2007) which monitored 29 dwellings in the United Kingdom (UK) during the summer of 2005–2006 and estimated mean indoor temperatures for living rooms to be 19.8 °C and 19.3 °C for bedrooms while Firth and Wright (2008) monitored 224 dwellings in the UK estimating mean living room temperatures at 21.4 °C and bedroom temperature at 21.5 °C. On the other hand, Mavrogianni et al. (2010) monitored 36 dwellings in London during the summer of 2009. Daytime means in living rooms rose above 28 °C in three dwellings out of the total 36 sampled while average indoor temperatures in 53% of living rooms were above indoor thermal comfort temperatures of 25 °C.

The human thermoregulatory mechanism endeavors to maintain a constant core temperature for the body, which commonly requires that the internal heat generated by metabolism be transferred through the skin and lungs to the surrounding environment (Robinson 2000) with the human body temperature being normally maintained at approximately 37 °C by the anterior hypothalamus through thermo-regulation (Hifumi et al. 2018). Heat-related illness develops when the pathological effects of heat load cannot be eliminated from the human body (Szekely et al. 2015) manifesting through excessive loss of water which can induce dehydration and salt depletion (Hifumi et al. 2018). In the event of exposure to extreme high indoor temperatures, removal of excess waste from the body is impeded triggering the core temperature to rise and physical health problems can begin (Robinson 2000). Healthy humans have sufficient heat regulatory which cope with increases in temperature up to a particular threshold; however, beyond a certain point the thermoregulatory system can collapse (Kovats and Hajat 2008).

Climate change has the potential to result in more heat-related illnesses as the mean global temperatures rise (White-Newsome et al. 2012). Occupants with pre-existing diseases, children, and the elderly face the greatest risk from extreme indoor heat (Wright et al. 2017). The WHO estimates the global burden of disease from climate change risk factors to have caused 160,000 premature deaths particularly from heatwaves and floods (Myers et al. 2011). The chronically ill, elderly, and children spend considerable time inside dwellings thus making them more vulnerable to indoor heat exposure (Smargiassi et al. 2008; White-Newsome et al. 2012). Past studies show that Southern Africa is expected to be a climate change hotspot (Hoegh-Guldberg et al. 2018). Most experimental indoor temperature monitoring studies have been done in the Global North hence the study sort to fill this knowledge gap on the threat of climate change on indoor thermal environments in Africa. Adaptation techniques are needed immediately for the housing sector to deal with the impacts of extreme heat on indoor environments (Kinnane et al. 2016). The purpose of the study was to characterize summer indoor thermal environments in low-cost housing units on the South African Lowveld. The chapter is structured as follows: section "Materials and Methods" describes the materials and methods, section "Results" presents results, section "Discussion" presents a discussion of the study, while section "Conclusion" presents conclusion and plans for future work.

Materials and Methods

This section outlines the materials and methods used during the study. A brief description of the study site is given followed by a highlight of the indoor sensors used to gather indoor and ambient data.

Description of the Study Site and Sampled Dwellings

Agincourt/Matsavana (24.8279S: 31.2197E) is a low-income residential settlement on the Lowveld valley and is located in the town of Bushbuckridge Local Municipality in Mpumalanga Province (Wittenberg and Collinson 2017). All sampled dwellings in the study were of a detached nature and were constructed as standalone structures. Dwellings selected were constructed from either hollow block or clay standard bricks or standard cement and sand bricks with no wall plastering and no ceilings and were free running without any mechanical indoor temperature mechanisms but rather depend on occupant ventilation behaviors. Figure 1 represents the study site.

Figure 1 shows the settlement of Agincourt located in the Bushbuckridge Local Municipality in the Lowveld region of South Africa.

Figure 2 shows images of sampled dwellings for the study. The majority of dwellings were detached, roofed from iron corrugated iron sheets, none plastered with no ceilings. The left image represents low-cost dwellings, the middle image

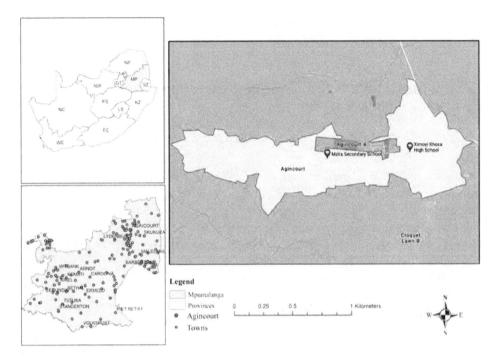

Fig. 1 Agincourt, Mpumalanga Province, South Africa

Fig. 2 Low-cost housing units sampled for the study

shows iButtons hanged inside a dwelling wall, and the right image shows the data downloading process.

Description of Indoor Temperature Monitoring Sensors: (*Thermochron iButton*)

Indoor air temperatures were measured using Thermochron iButton (DS1922L) manufactured by Maxim Integrated Products formerly Dalls Semiconductor (USA) as shown in Fig. 3. Thermochron iButton has a diameter of 17.35 mm and a thickness of 6 mm (Johnson et al. 2005). These sensors can measure air temperature range of −40 °C to 85 °C. Thermochron loggers are rugged, water-resistant, and self-sufficient sensors that measure temperature and record the result in a protected memory section (Hubbart et al. 2005). Figure 3 shows the image of a Thermochron *i*Button used in the study.

Indoor Air Temperature Measurements (T_{ai})

In each dwelling, two occupied spaces were identified: living rooms and bedrooms with a single sensor (Thermochron iButton) being installed in each room. Living rooms were chosen as this is the space where occupants spend considerable time during the daytime while the bedroom is mostly occupied at night. Air temperature in living rooms space was used as a proxy indicator or for occupant daytime exposure to extreme temperature while the bedroom air temperature was used as a proxy indicator for nighttime extreme indoor temperature exposure. Indoor sensors

Fig. 3 Thermochron *i*Button used for indoor and ambient temperature measurement

were placed at a standard height of 1.5 m–1.6 m above ground as applied in Healy and Clinch (2002), Yohanis and Mondol (2010), Newsome et al. (2012), Kane (2013), Lee and Lee (2015) and Magalhaes et al. (2016). Precautions were taken to make sure that sensors were not obstructed by furniture away from direct sunlight, heat, or any form of heat radiation (Malama and Sharpless 1996; Mavrogianni et al. 2010; Loughnan et al. 2015; Lee and Lee 2015; San Miguel-Bellod et al. 2018). Measurements were done continuously over a 24-h cycle at a 30-min temporal resolution.

Ambient Air Temperature Measurements (T_{air})

Ambient air temperature measurements were done using the same sensors as indoor temperatures (Thermochron *i*Button DS1922L). Similar to indoor measurements loggers were set to collect data at a 30-min temporal resolution over a 24-h cycle. Before deployment, all sensors were calibrated and tested for accuracy in the laboratory by the manufacturer (Matandirotya et al. 2019). Supplementary ambient air temperature data were obtained from the South African Weather Service (SAWS) nearest weather station.

Table 1 shows the number of summer indoor monitoring days during the 2016 and 2017 surveys. In 2016 indoor monitoring was done for a total of 17 days while for 2017 monitoring was done for 37 days.

Data Analysis

The study defined daytime as 8:00 am to 8:00 pm while nighttime was staggered from 9.00 pm to 7 am. For each room, hourly mean temperatures were calculated for the whole monitoring period. To establish the association/relationship between indoor and ambient temperatures at different times of day, indoor and ambient temperature were correlated using simple linear regression. Pearson values were used as a proxy indicator of insulation material strength. Section "Results" presents the results of the study.

Results

This section presents the results of the study. Tables 2 and 3 highlight the descriptive of indoor and ambient temperatures during the indoor monitoring campaigns.

Table 1 Indoor temperature monitoring times

Study site	From	To	Number of monitoring days
Agincourt 2016	13 April 2016	30 April 2016	17
Agincourt 2017	1 February 2017	9 March 2017	37

Table 2 Descriptive statistics for summer 2016

	Minimum (°C)	Maximum (°C)	Mean (°C)	Std. Deviation
Ambient daytime	17	38	27	5
Living room daytime ($n = 31$)	21	33	27	3
Bedroom daytime ($n = 19$)	20	34	27	2
Ambient nighttime	16	26	21	2
Living room nighttime ($n = 31$)	20	29	24	2
Bedroom nighttime ($n = 19$)	18	28	23	2

Table 3 Descriptive statistics for summer 2017

	Minimum (°C)	Maximum (°C)	Mean (°C)	Std. Deviation
Ambient daytime	19	41	29	7
Living room daytime ($n = 27$)	21	36	29	4
Bedroom daytime ($n = 27$)	21	37	29	4
Ambient nighttime	18	28	22	2
Living room nighttime ($n = 27$)	21	33	26	3
Bedroom nighttime ($n = 27$)	21	33	26	3

Table 2 represents descriptive statistics of indoor and ambient temperatures during the summer of 2016. The daytime ambient mean was higher than the nighttime mean by 6 °C. The trend was also observed for indoor temperatures during day and nighttime. The occupied spaces bedrooms and living rooms had similar means during daytime at 27 °C, while at nighttime there is 0.6 °C difference between living rooms and bedrooms with living rooms being slightly warmer. Living rooms are occupied spaces during daytimes so there was a potential of thermal discomfort as a result of temperatures exceeding the WHO maximum indoor temperature guideline of 24 °C. During the nighttime, there was also a marginal chance of thermal discomfort as the bedroom temperatures were close to breaching the 24 °C mark.

Table 3 represents descriptive statistics of indoor and ambient temperatures during summer 2017. A similar trend to 2016 was observed as mean daytime ambient temperatures were higher than nighttime temperatures by a 7.4 °C. Occupied spaces (living rooms and bedrooms) showed similar daytime and nighttime behaviors. Daytime differences between bedrooms and living rooms were by 0.2 °C, while at nighttime the difference was by 0.4 °C. There were very marginal differences between the rooms sampled. Figure 4 represents the density distribution of indoor temperatures in bedrooms during the summer of 2016.

Figure 4 represents indoor temperature density distribution in bedrooms during the summer of 2016. Throughout the summer monitoring period, all bedrooms had median temperatures above 24 °C beyond the WHO maximum temperature

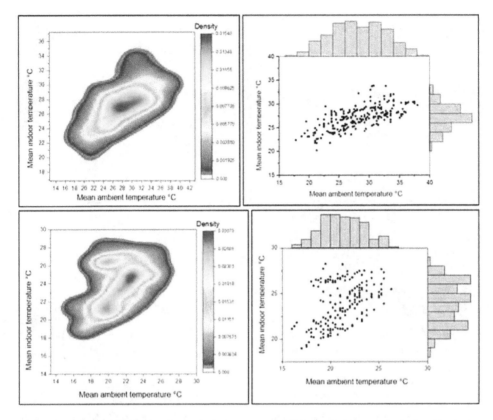

Fig. 4 Density distribution and marginal histograms of indoor temperatures in bedrooms during the summer of 2016. Top-row represents daytimes while the bottom row represents nighttimes

guideline for thermal discomfort. High bedroom temperature beyond the prescribed 24 °C has the potential to cause sleep impairment especially at night when space is used for night resting. Besides causing sleep impairment high indoor temperatures can cause excessive sweating that ultimately can cause dehydration as well as heat exhaustion. Figure 5 shows the density indoor temperature distribution for bedrooms during the summer of 2017.

Figure 5 shows indoor temperature density in bedrooms during the summer of 2017. During daytimes, indoor temperatures were in the region above 25 °C with minimum hourly mean temperatures being above 20 °C in all bedrooms sampled. The study assumed that this space is not occupied during the daytime; therefore, no risk was anticipated to occupants. According to Fig. 4, temperatures remained high at nighttime with minimum mean nighttime temperature remaining above 20 °C. The high indoor temperatures are mainly from solar radiation absorption which happens during the day which spills over into nightime since occupants do not have mechanisms to artificially regulate their indoor environments. The impact of these high nighttime indoor temperatures is that occupants are likely to suffer sleep impairment with a potential to loose fluids. With these high ambient temperatures, there is a likelihood of indoor overheating during both daytime and nighttimes.

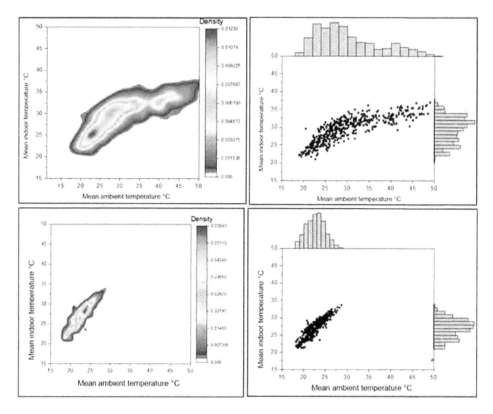

Fig. 5 Density distribution and marginal histograms of indoor temperatures in bedrooms during the summer of 2017. Top-row represents daytime while the bottom row represents nighttime

Figure 6 shows indoor temperature density distribution in living rooms during the summer of 2016.

Figure 6 shows the density distribution and marginal histograms of indoor temperatures in living rooms during the summer of 2016. The study assumed that living rooms are occupied in space during the daytime. The highest concentration was recorded at mean indoor temperatures of 26 °C during the day, while at night the highest concentration was at 24 °C. The biggest threat to human indoor thermal comfort came from daytime occupation of living rooms as temperatures were at the most time above 24 °C the prescribed thermal comfort temperatures by the WHO. The impact on occupants is that they are likely to lose a lot of bodily fluids from this exposure to high temperatures.

This trend was similar to that observed in bedrooms over the same summer monitoring period. A threat to human thermal comfort can only be experienced in this space during the day as people are expected to be occupying this space while on the other hand if the temperatures breach the 24 °C threshold at night it is a threat to those households that use living space for sleeping purposes thus sleep impairment can happen. Concerning thermal insulation material, the high indoor temperatures

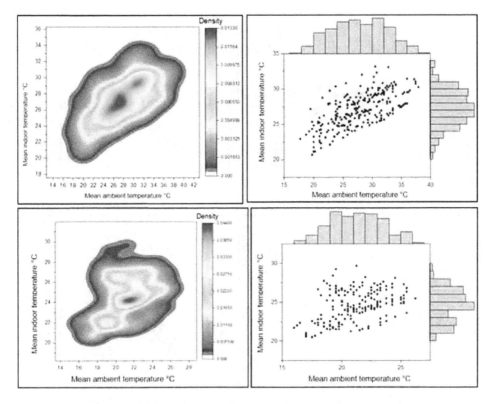

Fig. 6 Density distribution and marginal histograms of indoor temperatures in living rooms during the summer of 2016. Top row represents daytime while the bottom row represents nighttime

indicate that the thermal material is of poor thermal capacity. Figure 7 shows indoor temperature density distribution in living rooms during the summer of 2017.

Figure 7 shows the density distribution of indoor temperatures during the day and nighttimes. A similar trend was observed as in the summer of 2016 where the daytime minimum mean temperatures were recorded at above 20 °C while at nighttime they remained high with the minimum mean recorded in living rooms being at 21 °C. The impact on occupants was expected during the daytime as the study expected occupants to be occupying this space. Negative thermal effects are expected in living rooms if the indoor temperatures are above 24 °C during the day as people are expected to be occupying this space while it can have negative thermal impacts if occupants use this space for sleeping purposes during the night. The high indoor temperatures show that the thermal material used is weak in comparison to the high ambient temperatures experienced in the region thereby exposing occupants to extremely high indoor temperatures. It, therefore, implies that dwellings need to be constructed from a thermal material with the appropriate R-value and resistance capability to withstand the high levels of solar radiation. Measuring R-values of the building material was beyond the scope of this thesis but can be a future line of

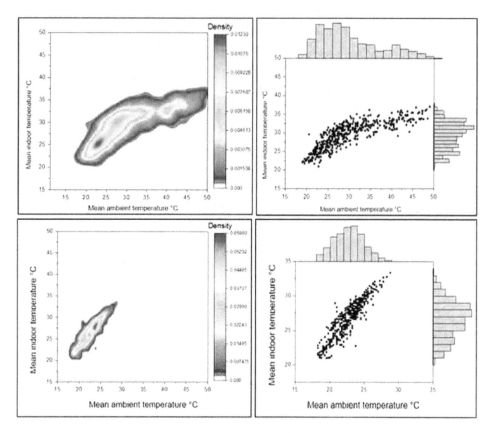

Fig. 7 Density distribution of indoor temperatures and marginal histograms in living rooms over the summer 2017 monitoring period. Top-row represents daytimes while the bottom row represents nighttimes

research. Figure 8 shows simple linear regression results for mean indoor and ambient temperatures in bedrooms during the summers of 2016 and 2017.

Figure 8 shows the simple linear regression results in bedrooms during the 2016 and 2017 indoor monitoring surveys. During both surveys, ambient temperatures were a good predictor of indoor temperatures with the strongest association being observed during the nighttime of 2017 monitoring (r = 0.90), while the least association was observed during nighttime of 2016 (r = 0.47). The strong relationship between ambient air temperatures and indoor temperatures shows the weakness in the thermal fabric of the material used to construct the sampled dwellings. Figure 9 shows simple linear regression results for mean indoor and ambient temperatures in living rooms during the summers of 2016 and 2017.

Figure 9 shows simple linear regression results of mean indoor and ambient temperatures in living rooms. The strongest association was observed during the night of 2017 monitoring (r = 0.90) while the least association was observed during nighttimes of 2016 at r = 0.40. The strong relationship indicates that these low-cost dwellings are highly sensitive to ambient temperature changes.

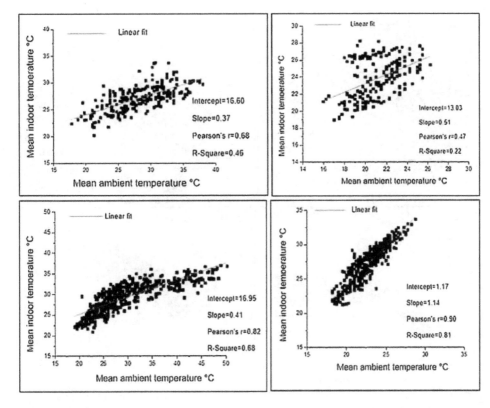

Fig. 8 Regression results for mean indoor and ambient temperatures in bedrooms. Top-row represents bedrooms 2016, while the bottom row represents 2017. The left image represents daytime while right image represents nighttime

Discussion

During summer 2016, the study observed that 100% of bedrooms had mean daytime temperatures above 24 °C through the monitoring period, while 53% of the same bedrooms recorded night temperatures exceeding 24 °C. In 2017, during both daytime and nighttime, 100% of bedrooms had daytime mean temperatures above 24 °C. The implication of being exposed to such indoor temperatures for a long time is that occupants lose a lot of fluids that can lead to dehydration. At night sleep impairment can also be experienced due to the high indoor temperatures. The population likely to be negatively affected most are young children and the elderly. Young children suffer from high heat exposure because their thermoregulatory system will not have developed much while for the elderly the thermoregulation system starts to suffer from dysfunction as sweating glands get blocked with aging hence not much sweat is generated to cool off the body. In both instances, it results in heat build-up within the body thereby putting a strain on the core. A strained core can end with such heat-related negative illnesses such as heat strokes, heat cramps, or heat exhaustion (Myers et al. 2011).

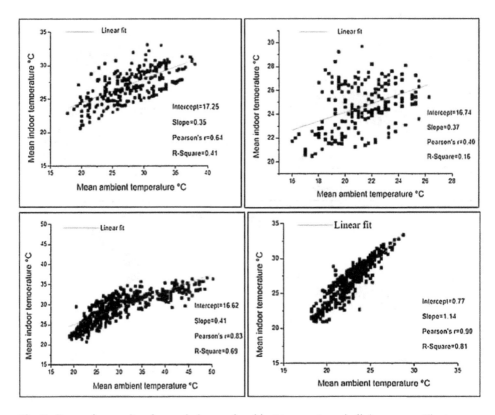

Fig. 9 Regression results of mean indoor and ambient temperatures in living rooms. The top row represents living rooms during 2016 while the bottom row represents living rooms 2017. The left image represents daytime, while right image represents nighttime during summer 2017

A similar trend to bedroom indoor temperature variations was observed in living rooms wherein 100% of living rooms in 2016 breached the 24 °C threshold, while at night 87% of sampled dwellings had mean temperatures above 24 °C. In 2017, the study observed that 100% of living rooms had mean daytime and nighttime indoor temperatures above 24 °C providing ideal conditions for indoor overheating. The high indoor temperatures recorded by the study during 2017 can be attributed to the monitoring period wherein temperature monitoring was done during the peak of summer season in February unlike the 2016 survey where temperature monitoring was done in April which is towards the period of transition to winter on the South African Lowveld. For living rooms, the study anticipated the greatest threat to indoor thermal comfort during daytime as people are occupying this space unlike at night where occupants shift to bedrooms. Prolonged exposure to high indoor temperatures can have devastating effects on those occupants already having underlying health conditions which can worsen.

The study also established that the sampled structures were highly sensitive to ambient temperature changes as confirmed by the strong positive Pearson correlations between indoor and ambient temperatures. The sensitiveness was observed for both day and nighttimes. This confirms that the thermal fabric of sampled dwellings is weak with an inability to regulate indoor temperatures as desired. If the thermal insulation is fully functional, it could act as a barrier to incoming solar radiation during the day and also be complemented by various ventilation practices put in place by occupants. Since the dwellings sampled were free running, occupants had no opportunity to artificially regulate indoor thermal environments. An increase in heat weather events in South Africa is likely to increase exposure to elevated temperatures for the poor which can be classified as a climate-related health threat (Wright et al. 2017). With ambient temperatures expected to continue on an upward trajectory as a result of climate change, there is a need to improve the thermal fabric of low-cost dwellings which can be achieved through thermal insulation retrofits for existing housing stock while for a new housing stock passive designs can be integrated. These two adaptation strategies can be of help to regulate indoor thermal conditions. The study findings concur to studies by Makaka and Meyer (2005), Naicker et al. (2017), Matandirotya et al. (2019) which also estimated that low-cost dwellings are sensitive to ambient temperature changes because of poor insulation. In a study by Mavrogianni et al. (2010), 53% of livingrooms average indoor temperatures were above thermal comfort levels of 24 °C while 86% of recorded temperatures could result in sleep impairment. Residents of low-cost dwellings are therefore at risk of being exposed to indoor thermal discomfort as temperatures continue to rise as a consequence of climate change.

Extreme high indoor temperature as a result of climate change is bringing several challenges for the built environment as people spend considerable time indoors; hence, the effects are likely to belong-lasting if appropriate adaptation interventions are not introduced especially for the poor marginalized populations who already occupy substandard housing. The study, therefore, brings to the fore the current existing indoor thermal environments in low-cost housing in the context of climate change. One of the possible immediate mitigation measures includes thermal insulation retrofits for the existing housing stock while for new housing stock passive designs can be incorporated. These measures improve the ability of housing structures to regulate indoor thermal environments. Furthermore, greening programs, urban planning, and housing are also other strategies that can be used to mitigate against devastating health threats from climate change (Wright et al. 2017). Future work will focus on measuring the thermal resistance capacity of building materials used for the construction of low-cost dwellings on the South Africa Lowveld region.

Study Limitations

The study had single sensors deployed in each room during monitoring; therefore, there was no mechanism to account for indoor vertical and horizontal temperature gradients. To mitigate this limitation, the study had to position sensors at the center of rooms where possible without interfering with occupants' daily activities. The study also could not take into account other indoor sources of radiant heat or ventilation practices that could impact on indoor temperatures. Additionally, the other limitation was that the study could not measure humidity which could facilitated the calculation of apparent temperatures which is an indicator of thermal sensation. Future studies will take into account these factors.

Conclusion

The study estimated that there is a risk of indoor overheating in low-cost dwellings on the South African Lowveld as a consequence of poorly insulated dwellings and a rise in temperatures from climate change. Occupants were exposed to indoor temperatures which breached the WHO maximum thermal guideline of 24 °C subjecting occupants to various physical health threats in the event of prolonged exposure. It is therefore imperative that adaptation and mitigation strategies on the existing housing stock are applied in order to reduce the effects of climate change on occupants. Future work will focus on community participation in the development of housing designs that are climate resilient and suit the changing climatic conditions of Southern Africa.

Acknowledgments Acknowledgments go to Brigitte Language who did data collection for the study working under the Prospective Household Observational cohort study of Influenza, Respiratory Syncytial Virus and other respiratory pathogens community burden and Transmission dynamics in South Africa (The PHIRST Study) supported by a grant from the United States Centre for Disease Control and Prevention and conducted by the National Institute for Communicable Disease of the National Health Laboratory Service. Ethical approval for the PHIRST study was obtained from the University of Witwatersrand, Johannesburg HREC (150808). Acknowledgments are also extended to South Weather Service for the provision of ambient temperature data.

References

Araos M, Berrang-Ford L, Ford JD, Austin SE, Biesbroek R, Lesnikowski A (2016) Climate change adaptation planning in large cities: a systematic global assessment. Environ Sci Pol 66:375–382. https://doi.org/10.1016/j.envsci.2016.06.009
Chersich MF, Wright CY, Venter F, Rees H, Scorgie F, Erasmus B (2018) Impacts of climate change on health and wellbeing in South Africa. Int J Environ Res Public Health. https://doi.org/10.3390/ijerph15091884
Dosio A (2016) Projection of temperature and heat waves for Africa with an ensemble of CORDEX regional climate models. https://doi.org/10.1007/s00382-016-3355-5
Firth SK, Wright AJ (2008) Investigating the thermal characteristics of English dwellings: summer temperatures: proceedings of conference: air conditioning and the low carbon cooling challenge: Cumberland Lodge, Windsor, 27–29 July 2008. Network for Comfort and Energy, London. Available at http://nceub.org.uk. Accessed 20 Jan 2019
Healy JD, Clinch JP (2002) Fuel poverty, thermal comfort and occupancy: results of a national household-survey in Ireland. Applied Energy 73(3–4):329–343. https://doi.org/10.1016/S0306-2619(02)00115-0
Hifumi T, Kondo Y, Shimuzu K, Miyake Y (2018) Heatstroke. J Intensive Care 6:30. https://doi.org/10.1186/s40560-018-0298-4
Hoegh-Guldberg O, Jacob D, Taylor M, Bindi S, Brown I, Camiloni A, Diedhiou R, Djalante K.L, Ebi F, Engelbrecht J, Guiot Y, Hijioka S, Mehrotra A, Payne S.I, Seneviratne A, Thomas Warren R, Zhou G (2018) Impacts of 1.5 °C of global warming on natural and human systems. In:

Masson-Delmotte V, Zhai P, Portner HO, Roberts D, Skea J, Shukla PR, Pirani A, Moufouma-Okia W, Pean C, Pidcock R, Connors S, Matthews JBR, Chen Y, Zhou X, Gomis MI, Lonnoy E, Maycock T, Tignor M, Waterfields T (eds) Global warming of 1.5 °C. An IPCC Special Report on the impacts of global warming of 1.5 °C above pre-industrial levels and related global greenhouse gas emission pathways, in the context of strengthening the global response to the threat of climate change, sustainable development, and efforts to eradicate poverty. [In Press]. https://www.ipcc.ch/sr15/chapter/spm/

Hubbart J, Link T, Campbell C, Cobos D (2005) Evaluation of a low-cost temperature measurement system for environmental applications. Hydrol Process 19:1517–1523. https://doi.org/10.1002/hyp.5861

IPCC (2007) Climate change 2007: the physical science basis: contribution of Working Group 1 to the Fourth Assessment Report of the Intergovernmental Panel on Climate Change. Cambridge University Press, Cambridge, UK/New York. https://wg1.ipcc.ch/publications/wg1-ar4/faq/docs/AR4WG1_FAQ-Brochure_LoRes.pdf

Johnson AN, Boer BR, Woessner WW, Stanford JA, Poole GC, Thomas SA, O'Daniel SJ (2005) Evaluation of an inexpensive small-diameter temperature logger for documenting ground water-river interactions. Groundwater Monitoring & Remediation 25(4):68–74. https://doi.org/10.1111/j.1745-6592.2005.00049.x

Kane T (2013) Indoor temperatures in UK dwellings: investigating heating practices using field survey data. Ph.D. thesis, Loughborough University Institutional Repository. https://hdl.handle.net/2134/12563

Kavgic M, Summerfield A, Mumovic D, Stevanovic ZM, Turanjanin V, Stevanovic ZZ (2012) Characteristics of indoor temperatures over winter for Belgrade urban dwellings: indications of thermal comfort and space heating energy demand. Energ Buildings 47:506–514. https://doi.org/10.1016/j.enbuild.2011.12.027

Kinnane O, Grey T, Dyer M (2016) Adaptable housing design for climate change adaptation. Proc Inst Civ Eng Eng Sustain 170(5):249–267. https://doi.org/10.1680/jensu.15.00029

Kovats SR, Hajat S (2008) Heat stress and public health: a critical review. Annu Rev Public Health 29:41–55. https://doi.org/10.1146/annurev.publhealth.29.020907.090843

Lee K, Lee D (2015) The relationship between indoor and outdoor temperature in two types of residence. Energy Procedia 78:2851–2856. 6th international building physics conference, IBPC 2015. https://doi.org/10.1016/j.egypro.2015.11.647

Loughnan M, Carrol M, Tapper NJ (2015) The relationship between housing and heatwave resilience in older people. Int J Biometeorol 59:1291–1298. https://doi.org/10.1007/s00484-014-0939-9

Magalhaes SMC, Leal VMC, Horta IM (2016) Predicting and characterizing indoor temperatures in residential buildings. Results from a monitoring campaign in Northern Portugal. Energ Buildings 119:293–308. https://doi.org/10.1016/j.enbuild.2016.03.064

Makaka G, Meyer E (2005) Temperature stability of traditional and low-cost modern housing in the Eastern Cape, South Africa. J Build Phys 30(1). https://doi.org/10.1177/1744259106065674

Malama A, Sharpless S (1996) Thermal performance of traditional and contemporary housing in cool-season of Zambia. Build Environ 32(1):69–78. https://doi.org/10.1016/S0360-1323(96)00036-4. 1997

Matandirotya NR, Cilliers DP, Burger RP, Language B, Pauw C, Piketh SJ (2019) The potential for domestic thermal insulation retrofits on the South African Highveld. Clean Air J 29(1). https://doi.org/10.17159/2410-972X/2019/v29n1a1

Mavrogianni A, Davie M, Wilkinson P, Pathan A (2010) London housing and climate change: impact on comfort and health-preliminary results of a summer overheating study. Open House Int 35(2):49–59

Miguel-Bellod JS, Gonzalez-Martinez P, Sanchez-Ostiz A (2018) The relationship between poverty and indoor temperatures in winter: determinants of cold homes in social housing contexts from 40s–80s in Northern Spain. Energ Buildings 173:428–442. https://doi.org/10.1016/j.enbuild.2018.05.022

Myers J, Young T, Galloway M, Manyike P, Tucker T (2011) A public health approach to the impact of climate change on health in Southern Africa-identifying priority modifiable risk. S Afr Med J

= Suid-Afrikaanse tydskrif vir geneeskunde 101(11):817–820. https://doi.org/10.7196/samj.5267

Naicker N, Teare J, Balakrishna Y, Wright CY, Mathee A (2017) Indoor temperatures in low-cost housing in Johannesburg. Int J Environ Res Public Health 14(11). https://doi.org/10.3390/ijerph14111410

Robinson PJ (2000) On the definition of a heat wave. J Appl Meteorol 40. https://doi.org/10.1175/1520-0450(2001)040<0762:OTDOAH>2.0.CO;2

Smargiassi A, Fourneir M, Griot C, Baudouin Y, Kosatsky T (2008) Prediction of the indoor temperatures of an urban area with an in-time regression mapping approach. J Expo Sci Environ Epidemiol 18:282–288. https://doi.org/10.1038/sj.jes.7500588

Summerfield AJ, Lowe RJ, Bruhns HR, Caeiro JA, Steadman JP, Oreszcyn T (2007) Milton Keynes Energy Park revisited: changes in internal temperatures and energy usage. Energ Buildings 39:783–791. https://doi.org/10.1016/j.enbuild.2007.02.012

Szekely M, Carletto L, Garami A (2015) The pathophysiology of heat exposure. Temperature 2 (4):452. https://doi.org/10.1080/23328940.2015.1051207

Webber T, Haensier A, Rechid D, Pfeifer S, Eggert B, Jacob D (2018) Analysing regional climate change in Africa in a 1.5,2 and 3 °C global warming world. Earth's Future 6:643–655. https://doi.org/10.1002/2017EF000714

White-Newsome JL, Sanchez BN, Jolliet O, Zhang Z, Parker EA, Dvonch JT, O'Neil MS (2012) Climate change and health: indoor heat exposure in vulnerable populations. Environ Res 112:20–27. https://doi.org/10.1016/j.envres.2011.10.008

Wittenberg M, Collinson M (2017) Household formation and household size in post-apartheid South Africa: evidence from the Agincourt sub-district 1992–2012. Demogr Res 37:Article 39, 1297–1326. https://doi.org/10.4054/DemRes.2017.37.39

Wright CY, Street RA, Cele N, Kunene Z, Balakrishna Y, Albers PN, Mathee A (2017) Indoor temperatures in patient waiting rooms in eight rural primary health care centres in Northern South Africa and the related potential risk to human health and wellbeing. Int J Environ Res Public Health 14(1):43. https://doi.org/10.3390/ijerph14010043

Yohanis Y, Mondol D (2010) Annual variations of temperature in a sample of UK dwellings. Appl Energy 87:681–690. https://doi.org/10.1016/j.apenergy.2009.08.003. (2009)

4

Triple Helix as a Strategic Tool to Fast-Track Climate Change Adaptation in Rural Kenya: Case Study of Marsabit County

Izael da Silva, Daniele Bricca, Andrea Micangeli, Davide Fioriti and Paolo Cherubini

Contents

Introduction ....
The Triple Helix
    The Role of the University
    The Relationships between University, Government, and Industry
    Triple Helix in Africa
    Triple Helix for Sustainability and Climate Change
Energy Access and Climate Change in Kenya
    Country Overview ...

    Rural Electrification in Kenya
Case Study – Marsabit County, Kenya ...
    Overview
    Energy Access in Marsabit ...

I. da Silva (✉)
Strathmore University, Nairobi, Kenya
e-mail: idasilva@strathmore.edu

D. Bricca
Sapienza University of Rome, Rome, Italy

A. Micangeli
DIMA, Sapienza University of Rome, Rome, Italy
e-mail: andrea.micangeli@uniroma1.it

D. Fioriti · P. Cherubini
DESTEC, University of Pisa, Pisa, Italy
e-mail: davide.fioriti@ing.unipi.it; paolo.cherubini@ing.unipi.it

Marsabit Town
Illaut
Laisamis
Discussion
Conclusions
References

Abstract

The lack of affordable, clean, and reliable energy in Africa's rural areas forces people to resort to poor quality energy source, which is detrimental to the people's health and prevents the economic development of communities. Moreover, access to safe water and food security are concerns closely linked to health issues and children malnourishment. Recent climate change due to global warming has worsened the already critical situation.

Electricity is well known to be an enabler of development as it allows the use of modern devices thus enabling the development of not only income-generating activities but also water pumping and food processing and conservation that can promote socioeconomic growth. However, all of this is difficult to achieve due to the lack of investors, local skills, awareness by the community, and often also government regulations.

All the above mentioned barriers to the uptake of electricity in rural Kenya could be solved by the coordinated effort of government, private sector, and academia, also referred to as Triple Helix, in which each entity may partially take the other's role. This chapter discretizes the above and shows how a specific county (Marsabit) has benefited from this triple intervention. Existing government policies and actions and programs led by nongovernmental organizations (NGOs) and international agencies are reviewed, highlighting the current interconnection and gaps in promoting integrated actions toward climate change adaptation and energy access.

Keywords

Triple helix · Rural electrification · Wealth creation · Climate change adaptation · Suitable policies and regulations

Introduction

Energy poverty is a serious concern in developing countries which often are the ones to bear most of the negative consequences of climate change. Currently, about 2.6 billion people have no access to clean cooking and about 850 million do not have access to electricity (International Energy Agency (IEA) 2019a). Some of these consequences are deficient healthcare systems, lack of access to education, and lower economic growth leading to political instability and migratory flows. Despite of international agencies' support and local government fund allocation, still 500 million people are expected to have no access to electricity by 2030 (International

Energy Agency (IEA) 2019b). All of the above are worsened by climate upheaval such as droughts, floods, and rain shortage (Trenberth et al. 2014).

Energy, food, and water are scarce and of low quality for people living in rural areas of developing countries. Food insecurity affects about 10–11% of people worldwide with peaks of 30% in east Africa (FAO 2019), or even beyond as confirmed by the surveys performed by Fraval et al. (2019). Food is often cooked without proper cooking facilities, often with firewood or charcoal, and exhaust gases are released inside the rooms without a proper ventilation system (Goldemberg et al. 2018). This affects the healthy conditions of inhabitants and especially children (Smith et al. 2014). Moreover, according to UNICEF, only about 30% of people in Sub-Saharan Africa had access to safely managed drinking water in 2017, whereas 30% of them had limited or inadequate access (UNICEF and WHO 2019). Often, water is far from settlements and unemployed people, usually children or women, have to walk several kilometers a day for procuring water for the family (UNICEF and WHO 2019). Furthermore, lack of water and scarce hygiene can facilitate the widespread of diseases (UNICEF and WHO 2019). Energy poverty, hunger, and no access to clean water are extremely correlated not only among themselves but also with low hygiene and low development rates (FAO 2019).

Electricity is widely recognized as a major determinant for social and economic development (Gambino et al. 2019), as it can allow the use of modern devices to be used for commercial or small industrial activities. The attention that African governments have paid on fostering access to electricity is justified by the correlation between electricity and income, as well as with the social development index (Zhang et al. 2011; FAO 2019). In fact, access to modern energy can have significant effects on poverty and access to water and reduce hunger, which are all significant priorities stated by United Nations (International Energy Agency (IEA) 2019a). Renewable energy–based electricity can provide reduction in CO_2 emissions by reducing the recourse to coal-based electricity in communities already reached by the national grid, by hybridizing with renewable sources existing mini grids mostly powered by diesel gensets (Micangeli et al. 2017), and by displacing the use of traditional fuels for lighting (Micangeli et al. 2017; Ambition to Action 2019). As an example, small-scale irrigation can contribute to income diversification and livelihood resilience (Murphy and Corbyn 2013).

Governments in developing countries often cannot enforce a fast development of rural electrification alone, due to lack in financial resources, know-how, or human capacity (Franz et al. 2014). On this regard, the involvement of the private sector as well as know-how holders, such as universities and international entities, is considered crucial (Franz et al. 2014), also to prevent the consequences that climate change may cause on already critical situations. Recent modern business models exploit the benefits of the integrated approach of the so-called energy-water-food nexus, by which the developer brings in electricity service, which in turn avails services like water and food through irrigation (Res4Africa 2019).

Universities produce knowledge and promote long-term growth through innovation and technology transfer to private and public sector. On the other hand, governments develop guidelines and policies in order to guide the industry toward the long-term social benefits for society. By considering the above, it is clear that significant synergies among the three entities can be explored and this cooperation is called "Triple Helix" (Leydesdorff and Etzkowitz 1995; Carayannis and Campbell 2010), which is proposed in this chapter as a tool to foster climate change adaptation in Kenya, with the special attention to the Marsabit County, an area heavily hit by climate change and subject to draughts and floods.

In section "The Triple Helix," the Triple Helix concept is detailed, including applications for rural electrification, Africa, and sustainability concerns. In sections "Energy Access and Climate Change in Kenya" and "Case Study – Marsabit County, Kenya," the case study is introduced and in section "Discussion" the major discussions are proposed. Finally, the conclusions are made.

The Triple Helix

The concept of Triple Helix (TH) has been proposed in the 1990s (Leydesdorff and Etzkowitz 1995) as a new framework to conceptualize relationships between university, government, and industry, to address the need of organizational innovation in a knowledge-based society. Innovation, defined as "reconfiguration of elements into a more productive combination" (Etzkowitz 2008), requires, in this view, a central role of the university, since innovation is not merely attributable to new product development by the industrial sector (Etzkowitz 2008).

The Role of the University

It has been acknowledged, in a seminal work made by Gibbons et al. (Gibbons et al. 1994), that the very production of knowledge has transitioned from a "traditional" framework, labeled Mode 1, driven and administered by classic academia in a disciplinary context, to a new mode, labeled Mode 2. In this new framework, knowledge is produced in the context of application, by a variety of actors coming from different disciplines, in an interactive relationship with society as a whole, both in defining the problems and research goals, and in diffusion of results. In recognition of the different way of knowledge production, and the role that it acquired in innovation, the university reshaped its mandate. The first modern universities were created in the twelfth century with the purpose of preservation and dissemination of knowledge (teaching mission). In the nineteenth century a second mission – research – was added, and more recently a third mission, that of contribution to economic development (Etzkowitz 2003). This mission requires an entrepreneurial model to be adopted by academia, that is not limited to preservation and transmission of knowledge but is involved proactively in technology transfer and incubation activities (Etzkowitz 2003). Furthermore, Carayannis et al. proposed the concept of "Mode 3" in which knowledge is produced by pluralism and diversity of knowledge, in a Triple, Quadruple, or Quintuple Helix framework, also including a combination of modes 1 and 2 (Carayannis and Campbell 2010).

The Relationships between University, Government, and Industry

The relationships among these three entities can be visualized as lying among two extremes: a centralized model where the government guides science policy and industrial development ("statist model") or a situation of very limited interaction, where each entity attains its core mandate separately from the others ("laissez-faire model") (Etzkowitz 2008). To visualize the interactions typical of the TH, Fig. 1 shows the "field interaction model" proposed by Eskowitz (Etzkowitz 2008) that schematizes government, industry, and university as three spheres with an inner core, representing their specific functions, and outer cores that overlap as each entity "takes the role of the other on a synergetic manner."

It is argued that the TH concept could act as a flywheel in pursuing a path of compliance with the objectives set out by the Sustainable Development Goals (SDGs) for rural areas of developing countries. However, TH dynamics has usually referred to a context of discontinuous innovation in advanced technological sectors such as ICT (Information and Communication Technology), semiconductors, or biotechnology. The most iconic example of regional development led by TH dynamics is the one of Silicon Valley, an international hub for high-tech companies animated by Stanford University's knowledge development (Etzkowitz 2012). In the following section, we will look at how the TH framework has been applied and discussed in the African context and how it has been integrated or reformulated to embed the sustainability issues underpinned by the SDGs.

Triple Helix in Africa

Rarely TH applications have been found in Africa. As noted by Outamha and Belhcen (Outamha and Belhcen 2020), the core mission of university in the

Fig. 1 Triple Helix field interaction model (Etzkowitz 2008)

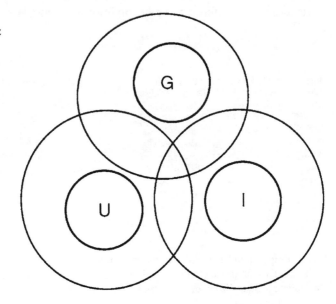

continent is still the provision of human capital to industry and government, hence it is not on par with the global standards of knowledge-based societies. The historical reasons for this have been identified by some authors in the focus on mass primary and universal education over higher education in aid programs, at the expense of higher education (Dzisah 2011).

In Outamha and Belhcen (2020), a list of challenges that interconnections between university and industry are facing in Africa is given and shown in Fig. 2, together with the challenges for the implementation of mini-grids for rural electrification (African Development Bank Group et al. 2016). The different points are shown below:

- The lack of data regarding mini-grid projects reflects the lack of interest and commitment from industry to share information and collaborate in knowledge creation and diffusion, aggravated by confidentiality and property right concerns.
- Lack of capacity for mini-grid development reflects the lack of interest and lack of communication platforms to socially disseminate results of application-oriented curricula (Perez-Arriaga et al. 2018).
- Lack of access to finance reflects the lack of entrepreneurial prowess on the university side, resulting in lack of interest and aggravated by lack of mutual trust and lack of communication platforms bringing together the other two sectors.
- Gaps in policy and regulation reflect that academic curriculums in universities, prioritizing human capital formation, tend to not address the training of leaders that can give an answer to structural problems undermining the development of countries.
- Lack of viable business models is also a consequence of a lack of coordinated effort from governments planning for electrification plans, state-owned utilities, private actors, and academics in analyzing and reviewing current on-field experiences and contributing to develop improved models, overcoming contradictions

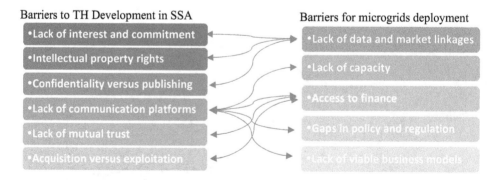

Fig. 2 Proposed linkages between barriers to university-industry linkages in SSA (Outamha and Belhcen 2020), on the left, and the barriers to development of Green Mini-Grids in SSA (African Development Bank Group et al. 2016), on the right

between knowledge acquisition by universities and commercial exploitation by private sector players.

The adoption of a TH framework has been proposed for the energy sector in SSA, specifically for the development of photovoltaic installations in Uganda (Da Silva and Wassler 2011). There are examples of spin-off companies operating in the Solar Home System sector in SSA, such as BBOXX (from Imperial College of London) and Azuri Technologies (from Cambridge University) (Bagley et al. 2018). By looking at these two companies, it is worth highlighting the issue of the "localization" of TH dynamics for the specific case, as they are firms born from non-African universities but working on African ground with local government and international entities.

In tackling the issue of electricity access conjointly with the food and water security and climate change adaptation issues, it has to be kept in mind that is a sector of intervention with serious concerns of bankability of investments (Kottász and Draeck 2017), especially for the realization of mini-grids, and in which international organizations and NGOs play a leading role. Therefore, there is, on the one hand, a multiplicity of stakeholders (i.e., NGOs) that are not present in the classic TH model, and on the other hand, a geographical globalization by which interventions directed at rural communities of developing countries involve not only local governments, universities, and industries but also international counterparts, mobilized by the concerns and goals of international agencies and non-profit organizations. As an example, in Moore et al. (2015) a program for ICT (Information and Communication Technology) education for Rwandan teachers is discussed as a possible example of TH development, albeit "loosely" not only because of the leading role of the local government in proposing the program, supported by local academia and privates, but also for the role of UNESCO and foreign cooperation agencies, factors that "are missing from the model."

Triple Helix for Sustainability and Climate Change

The original TH model did not entail sustainability issues linked to innovations. Etzkowitz and Zhou (Etzkowitz and Zhou 2006) proposed to consider a second TH system, one of university, government, and public to act as a counterbalance to technological innovations introduced by the university-government-industry triad, by representing the "dynamic of controversies over technological innovation." This picture of twin helices was chosen to maintain the triadic nature of each one, whereas other authors have proposed to the inclusion of a fourth helix – representing civil society, in form of NGOs or other associations of citizens – and a fifth helix, accounting for the natural environment of society (Carayannis et al. 2019). A quadruple helix approach has been proposed specifically for Africa (Kimatu 2015), as it could provide the needed "sustainable innovation ecosystem" to accelerate economic development. Similarly, a quadruple helix approach is advocated in (Amaro et al. 2014) for Guatemala, a country vulnerable to extreme climate events that shows a lack of awareness for climate change effects and related action systems, and that could therefore benefit from a TH intervention integrated with civil society actors.

Finally, other authors criticized the underlying paradigm of the university role in a TH model, namely the concept of third mission pursued by an entrepreneurial university, highlighting how the adoption of market logics promoted profit over the technology transfer; the new mission for academic institutions should be that of "creating societal transformations in pursuit of realizing sustainable development" (Trencher et al. 2014). In this perspective, with the goal of materializing sustainable development in a specific context, a function of "co-creation for sustainability" is proposed for the university, beyond the conventional research and development and product improvement through innovation.

Energy Access and Climate Change in Kenya

Country Overview

Kenya is an equatorial country of about 580,000 km^2 located in East Africa with significant different climate conditions varying from desertic to humid areas. With a gross domestic product (GDP) of 87.9 billion USD in 2018 with an average growth rate of about 9% in the past 10 years and a population of about 51.4 million people in 2018, Kenya is experiencing a strong economic growth (World Bank 2020). In the last 20 years, the GDP per capita has grown steadily and even poor people have benefited as the poverty ratio decreased by 36% over the period.

However, the Sustainable Development is not achieved in Kenya; still many people live below the poverty line, rarely water is safely managed with even no basic service in 40% of cases (UNICEF and WHO 2019), hunger significantly affects 10–40% of the children 5 year old or younger (FAO 2019), and sanitation and hygiene is even worse, also affecting maternity-related concerns (UNICEF and WHO 2019). Energy access is still limited: about 25% of people do not have access to electricity and an even larger proportion (75%) have no clean cooking facilities (World Bank 2020). Altogether, inequalities are significant, as also proven by the high 40% Gini index (2015) (World Bank 2020). Nevertheless, these problems have been decreasing.

In order to promote growth and tackle these challenges, the government has developed and implemented the Kenya Vision 2030 program in 2007 (Ministry of Planning and National Development 2007) with significant pillars based on macro-economic stability, measures for the poor, capacity building, research, infrastructures, and energy. A new constitution was promulgated in 2010 and the corresponding legislation and administrative procedures have been set up the following year. The reform laid the bases for the decentralization of power into county-level governments able to develop local policies and tax reform accounting for the specific problems of the territory (Ministry of Planning and National Development 2007; Ngigi and Busolo 2019). The Devolution, which was ushered in 2013, has benefits in terms of a more equitable distribution of resources, economic and social development, political participation of citizens, enhanced transparency and accountability of governments, and even unity (Ngigi and Busolo 2019).

Climate Change

Kenya is vulnerable to climatic events, mainly droughts and floods. According to the National Adaptation Plan (NAP) launched in 2016 (Government of Kenya 2016), droughts have been recurring in the past 40 years, affecting millions of people by disrupting pastoral and agricultural activities, causing famine and population displacement. Floods, which are recurring phenomena as well, cause loss of human lives and damages to infrastructure, such as roads or power lines, hence potentially leading to problems to the transportation sector and the supplying of basic services to the population. Moreover, the areas subject to floods can experience the spreading of waterborne or sanitation-related diseases in the subsequent period (Government of Kenya 2016), as humid conditions ease the widespread of diseases, excrements and sewage are difficult to contain, especially in case of open defecation (UNICEF and WHO 2019), and the damaged infrastructure delay possible aid (Bündnis Entwicklung Hilft (Alliance Development Works) and United Nations University – Institute for Environment and Human Security (UNU-EHS) 2016). Given the damage and risks due to these extreme events, local and international entities have developed policies and dedicated programs for droughts and disaster risk management in Kenya, as detailed in Global Water Partnership Easter Africa (GWPEA) (2015).

Rural Electrification in Kenya

In Kenya Vision 2030, the "long-term development blueprint for the country" launched in 2007 (Ministry of Planning and National Development 2007), energy is considered a foundational infrastructure to enable the goals set out by the government on the economic, social, and political governance pillars. To address the specific issue of electricity access in rural areas, the Rural Electrification Authority (REA) was created in 2007, and then with the 2019 Energy Act adopted in March 2019 (Republic of Kenya 2019), REA has been changed to Rural Electrification and Renewable Energy Corporation (REREC). REREC has the mandate to develop and implement the rural electrification program of Kenya and to manage a dedicated Rural Electrification Programme Fund, also established by the Energy Act.

This choice reflects the change of pace in the country's objectives: the target for 100% electricity access in rural areas originally set out for 2030 (Rural Electrification Authority (REA) 2008) has been moved to 2022 in 2013 (Kenya Ministry of Energy 2018), as electricity is considered a key backbone for the achievement of the "big four agenda" (Republic of Kenya – The National Treasury and Planning 2018) pivoted on enhancing food security, universal healthcare, affordable housing, and increasing the GDP share of manufacturing and agri-processing activities. Kenya has been the fastest African country in closing the access gap, with an annualized increase in access of 6.4 percent points between 2010 and 2017, with a remaining deficit of 18 million people without access, corresponding to a 63.8% access rate in 2017 (IEA et al. 2019).

According to the national electrification strategy, universal access will be achieved mostly via grid intensification and densification (2.77 million connections), standalone solar systems (1.96 million connections), grid expansion (269,000 connections), and mini-grids (35,000 connections) (Kenya Ministry of Energy 2018).

Research showed that the mere presence of grid infrastructure in Kenya does not translate to high rate of consumer takeoff, even for "under grid" households that are within 200 m of a low voltage power line (Lee et al. 2014). To address this issue, the Last Mile Connectivity Program has been launched, to tackle the key issue of affordability of the connection fee, lowered from 398 to 171 USD, including the opportunity of paying in monthly installments.

On the other hand, the construction of the mini-grids, foreseen by the national electrification strategy, will fall under the Kenya Off-Grid Solar Access Project (KOSAP), started by the Ministry of Energy in July 2017 and financed by the World Bank (Ministry of Energy 2018). The project benefits 14 underserved counties in the northern and eastern regions of Kenya, including the Marsabit County, and envisages providing standalone systems, for dispersed users and public facilities, clean cookstoves, solar water pumps for community facilities, and mini-grids in selected target communities. As per government data, the number grew to 151 after the field surveys (Ministry of Energy 2019a). It is worthy noticing that KOSAP integrates environmental concerns, citizen engagement, a capacity building component – also directed at county governments – and incentives for private sector involvement. The mini-grids to be developed under this framework, in fact, will be developed under a public-private partnership (PPP) lasting 7–10 years by which generation facilities will be co-funded by public and private actors, and distribution networks will be built with public funds only (KP and REA 2017). Until the PPP lasts, the O&M of generation and distribution assets will be performed by a single private service provider (PSP), while the final users will be KP customers subject to the national tariff (KP and REA 2017); thereafter, all the assets, hence both the generation and distribution ones, will be owned by the Government of Kenya.

According to the National Electrification Strategy, the funding allocated on improving the national grid are about 2.3 billion USD over five years, accounting for about 82% of the total. The interventions are expected to focus on the installation of new transformers (grid densification), on the extension of existing MV lines up to 2 km to reach new customers (grid intensification) and also on the extension of the distribution system with longer lines expansion within 15 km the existing backbone (grid extension). These measures focus on areas relatively close to the existing network; yet, rural areas are not overlooked. In fact, about 18% of the funding will fund mini-grids and solar home systems (SHS) with the goal of reaching the people that cannot be reached by the other measures. SHS will receive more than ten times the funding with respect to mini-grids (Kenya Ministry of Energy 2018); yet, the potential for mini-grid development in Kenya may have been underestimated in the national planning strategy, as detailed in a series of technical reports and scientific papers (Ambition to Action 2019).

Case Study – Marsabit County, Kenya

In this section we describe the major activities in the view of the TH concept that have been developed in Marsabit, Kenya. The information reported here are based on the available literature and the knowledge of the experienced authors, also based on field work. Three case studies have been chosen for this book chapter due to the different types of conditions and information they represent.

Overview

Marsabit is one of the 47 counties created with the 2010 Constitution of Kenya, located in the northern part of the country, bordering Ethiopia. According to the most recent census, held in 2019, Marsabit has 459,785 inhabitants occupying a total of 67,000 square kilometers, making it the least densely inhabited county in Kenya, with an average population density of 6 persons/km^2 (Kenya National Bureau of Statistics 2019). The agroclimatic zones range between very-arid (zone VII) to semi-humid/semi-arid (zone IV). The prevalent land cover is barren land (65.4% of the total area), which is only fit for pastoralist activities (Wiesmann et al. 2016). Rural population engaged in agri-pastoral activities is largely nomadic, moving around the countryside according to the seasons in search of food, water, and grazing land.

The poverty rate in Marsabit registered with the 2009 census was 75.8%, the fourth overall highest among Kenya, with a poverty severity of 8.8% with respect to a 4.89% national average (Wiesmann et al. 2016). The living conditions are very different among the various sub-counties and among towns and rural areas. All the settlements around the main northern road, which goes from Merile to Moyale and connects Kenya to Ethiopia, have the advantage of increased access to goods coming from all the county and benefit from the presence of numerous investors, which started setting foot in the region since devolution and significantly increased their presence during the last few years. In a comparative analysis done among Isiolo, Marsabit, and Meru counties, Marsabit exhibited the lowest resilience capacity index (FAO 2017). The main reason was a poor adaptive capacity, linked to low education levels and low income diversification, given the predominance of pastoral activities (FAO 2017).

Although the region is one of the lowest developed in Kenya, food insecurity affects a limited share of the population (10–15%) given the large consumption of milk, but there is serious/critical malnourishment among above 15% of children and it is also stated that the whole population is even at significant risk of food insecurity in case of drought emergency (UNICEF 2017). Moreover, despite improvement in the years, water access is still a problem. Socioeconomic analyses revealed that only 26.4% of population above 3 year old is currently enrolled in school, whereas 63.4%

of the population has never attended one (Kenya National Bureau of Statistics 2019). The typical economic activities are farming and livestock production, with a minority practicing fishing (1295 households out of 77,495), yet with no commercial-oriented business as the majority of households conduct these activities for subsistence (Kenya National Bureau of Statistics 2019). At the time of writing, the release of the 2019 census is still incomplete, so for other relevant data figures provided will refer directly or indirectly through secondary literature, to the 2009 census.

Given the climate and soil conditions, about 80% of the population is involved in pastoral production, a sector extremely vulnerable to not only recurrent droughts that have caused losses for over 60% of livestock in recent years but also floods that have also caused sweeping the cattle and killing them (The Ministry of Agriculture Livestock and Fisheries (MoALF) 2017). There are no perennial rivers but only four seasonal drainage systems (Marsabit County Government 2018); therefore, only 38.2% of the households have access to safe water (Wiesmann et al. 2016). The county is dependent on short rains for 80% of food security and on long rains for 20% (Kenya Food Security Steering Group (KFSSG) and Marsabit County Steering Group 2017). Moreover, the irrigated areas for the major crops (maize, tomatoes, and kales) declined also due to high maintenance costs for irrigation and low recharge of surface water sources (Kenya Food Security Steering Group (KFSSG) and Marsabit County Steering Group 2017).

Aiming to adapt to these challenges, some locals have attempted to develop activities such as water harvesting, value addition to local products, conservation of soil and water, tree planting, business diversification, selection of more resistant livestock types, yet in an insufficient and uncoordinated way, and the lack of access to electricity and good infrastructure are clearly additional barriers (The Ministry of Agriculture Livestock and Fisheries (MoALF) 2017). Deforestation in the lowlands in combination with droughts contributes to reduced water flows, health issues, and undermining of food security and strongly affects not only the living condition of the population but the farming businesses, agriculture targeting programs, and pastoral-ist pathways and lowers drastically the level of security due to land use conflicts (The Ministry of Agriculture Livestock and Fisheries (MoALF) 2017). Armed fights, which can also have a tribal nature across counties, result in human and livestock deaths, destruction of crops and homesteads, fear, and poverty.

Energy Access in Marsabit

Energy access in Marsabit County is restricted to urban centers, always supplied by off-grid solutions with the exception of the towns of Sololo and Moyale, which are interconnected to the Ethiopian national grid, thanks to their proximity to the border (Republic of Kenya and County Government of Marsabit 2018). A list of public mini-grids active in the county is provided in Table 1, the only known active private mini-grid in Table 2, and the list of 13 sites in which mini-grids will be built under KOSAP in Table 3. Also, a list of 26 medical, school, and public facilities to be

Table 1 Operational public mini-grids in Marsabit (Nygaard et al. 2018; Ambition to Action 2019)

Off-grid station	Diesel (kW)	Solar (kWp)	Wind (kWp)	Battery capacity (Ah)	Connections	Commissioning date
Marsabit	2900	0	500	0	8200 (June 2016)	1977
North Horr	184	0		0	160 (June 2016)	2016
Laisamis	264	80		27,816	160 (June 2016)	2016
Illaut	50	60	0	76,800	200	2019
Ambalo	50	60	0	76,800	200	2019
Balesa	50	60	0	76,800	n/a	2019
Total	3498	260	500	258,216	8920	

Table 2 Operational private mini-grids in Marsabit (Blodgett et al. 2017; Nygaard et al. 2018)

Mini-grid	Developer	Solar (Wp)	Commissioning date	Customers
Merile	Vulcan Inc.	1.5	2014	27

Table 3 Mini-grids to be built in Marsabit County under KOSAP (Ministry of Energy 2019a)

Mini-Grid	Potential customers
Bubisa	329
Kargi	379
South Horr	496
Loiyangalani	1051
El Gadhe	108
Gatab	299

benefited by KOSAP intervention in the county has been published (Ministry of Energy 2019b). Finally, there are also private projects, however, often supported by international cooperation (KfW/GIZ/EnDEV) such as for the mini-grids to be developed in Dukana, Illeret, Ngurunit, Dabel, and Diridima (Energies Development 2018; Nygaard et al. 2018).

Moreover, although being a low developed area, there have been developed some large-scale projects in recent years. First of all, given the large wind power potential available in areas of the region, a large scale 310 MW wind farm, representing about 15% of the installed capacity, has been developed close to lake Turkana and connected to the national grid in September 2018 (LTWP 2018). Furthermore, given the proximity with the Ethiopian border, a 263-km HVDC transmission line has been developed as part of the Eastern Electricity Highway Project interconnecting Kenya and Ethiopia (KETRACO 2018). Despite the location of these large electrification projects, the County has not benefited from them in terms of increased access to electricity because low energy demand does not justify the installation of substations. However, the projects have led to enhance and tarmac the 195-km road infrastructures connecting South Horr with Laisamis (code D371) and with Loiyangalani (code C77), with the objective of easing the transport of materials and equipment for the construction of the wind farm (Lake Turkana Wind Power Limited 2011). Moreover, under the LAPSSET project the tarmacking of the Isiolo-Moyale highway has been completed, with significant economic benefits and enhanced accessibility for the region (Republic of Kenya and County Government of Marsabit 2018). The next step of the project, to be developed in future years, is the development to complete the corridor with a railway line (Republic of Kenya and County Government of Marsabit 2018).

The increased level of economic activities along the Isiolo-Moyale road is also visible as shops and residential houses are quickly sprouting by the highway. Local people are moving slowly from pastoralism to different types of business in this area or are using part of their herd to try and invest in diverse activities.

The Marsabit County set out the development of renewable energy sources as a recommendation for climate change mitigation and adaptation, aspiring to exploit the solar and wind potential of the region (Republic of Kenya and County Government of Marsabit 2015). However, the same document stated the lack of human capacity and investment opportunities as barriers for development of energy projects, as well as high fuel costs in construction machineries that hinder investments (Republic of Kenya and County Government of Marsabit 2015).

During the 2013–2017 period, the county provided over 2000 energy saving charcoal burners (*jikos*) to households, 417 solar street poles, and distributed solar panels to schools with the goal of improving the household reliance on biomass and kerosene that can significantly affect the health of people (Republic of Kenya and County Government of Marsabit 2018). While the goal for solar street poles set out in the previous integrated development plan has been exceeded, the other targets for the distribution of 500,000 jikos, start-up of a solar equipment enterprise, installation of solar pump driven boreholes, and construction of a power station per constituency have not been fulfilled yet (Republic of Kenya and County Government of Marsabit 2015, 2018).

In the current development plan, not only the previous targets have been restated and expanded but also specific funds have been dedicated for Centre for Renewable Energy Studies and Research of Marsabit County (Republic of Kenya and County Government of Marsabit 2018) to help overcome the knowledge gaps limiting the supply of energy to the population.

Marsabit Town

Marsabit Town is the main town in the county and is situated on an isolated extinct volcano, Mount Marsabit, where the environmental conditions are completely different from the rest of the region.

Devolution, started in 2013, had a broad impact on the development of Marsabit Town that became the county government headquarter, thus, not only the center of all international organizations operating in the county but even of all the various ministries of regional government. As a result, the town is attracting a growing number of investors with several companies setting up business and effectively opening up the northern frontier region market. Marsabit market, especially during Saturdays, became the first market in the County. Cereals and legumes are imported directly from Ethiopia, while vegetables come from the area around Meru. Between 2013 and the 2019 more than 40 schools were built and one level 5 hospital, in addition to the one run by CARITAS. In the town there is the presence of numerous clinics and dispensaries, mostly in the scattered sublocations. Marsabit Town currently has more than 50 primary schools, 15 secondary ones, and 2 technical institutes, accounting both public and private institutions. All the secondary schools and most of the primaries are electrified, reflecting as per the first phase of REA's strategic plan for 2008–2012 (Rural Electrification Authority (REA) 2008).

The town is served by a mini-grid installed by REREC and operated by Kenya Power with an installed capacity of 3.1 MW (2.6 MW from three diesel generators and 500 kW from wind turbines). The system was commissioned in 1977, with the generation capacity growing steadily and number of connections steeply rising, especially after devolution came into place. As most of the public infrastructure has then been built and reaching around 8200 customers as per June 2016. The medium voltage lines used to connect the public facilities have then been used to give access to households located within 600 meters of an existing transformer, with the uptake from final users facilitated by the Last Mile Connectivity Program which lowered the upfront cost of connections. Consequently, most of the connected households are either located in the town center where most of the business activities take place, or in the vicinity of schools, hospitals, or dispensaries and county government buildings, but also the areas surrounding the town are slowly benefiting from the ramp-up in electricity access.

Water access is the primary issue of the population residing in the town. The water table is very deep, due to the characteristics of the volcanic soil, and the main source of water are shallow wells in the forests, water boozer coming from boreholes in the nearby settlements or through water harvesting.

National government programs focusing on floods and droughts resilience were already deployed before devolution, but through the county government presence in the town, the solutions have been increasingly adapted to the specific context collaborating directly with the NGOs operating locally, local associations and cooperatives. Numerous projects and training programs have been carried out on water harvesting, water management, and small-scale agriculture as focused programs on seeds restocking and drought-resistant crops.

Illaut

Illaut is a town located in the south-western part of Marsabit, near the border with Samburu county. Illaut, after the rehabilitation of the existing Laisamis–Ngurunit–Ilaut–South Horr–Loiyangalani road, funded by the African Development Bank and the Government of the Netherlands to connect the Wind Farm developed on the shore of Lake Turkana, became a central point of connection for all the surrounding areas. The chief of Illaut is responsible for everyone living in all the nearby villages.

In consideration of its relevance, Illaut has been selected as one of 26 sites for the installation of hybrid mini-grids by REREC (REREC 2019). The system was commissioned in October 2019, with around 20 customers connected (two shops, one church, and one dispensary plus households). Due to the lack of Internet connectivity, the smart-metering system is still not operating at the present time, while power generation and consumption is steady. Having network access is in fact mandatory for the KP metering systems, as to collect the revenues, monitor consumption, and bill collection is a big hurdle for most of the new hybrid mini-grids operated by Kenya Power. However, Safaricom started building a repeater tower in the end of 2019, which is supposed to start working during spring 2020, to grant network access to the Illaut and the nearby villages. Access to the network has been proven to bring direct economic and social benefits for everyone: mobile and Internet services have a transformational impact, offering life-enhancing financial and health services, as well as the simple ability to enhance communication within families and within the whole community.

Most of the houses are made out of bricks and the Tuesday market became the main trading post for goods coming from Meru and Nairobi, through Marsabit and Nyeri, and for local handmade products. The presence of a nearby shallow well provides access to water that has to be treated to be drinkable. This is not perennial and depends on rainfall, making it hard for people to access water during the long-lasting dry season.

Firewood is the primary source for cooking, followed by charcoal. In Illaut town local government is creating awareness regarding the negative that open fire has on the health of mainly women and children.

In the end of 2019 the county government, following the access to water plan of the Water Resources Management Authority (WARMA), started building a borehole in the town that will be completed in early 2020. Thanks to this borehole the town, and the nearby settlements, can have access to clean water that can be used for irrigation and cattle survival during the dry season. The chief of the village, together with key personalities of the community, has been engaging both the farmer communities in Marsabit and Laisamis, NGOs like PACIDA, and some SACCOs to obtain training, seedlings, and agriculture machines to open up the possibility of unlocking the potential of local farmers.

Illaut can be considered as an example of sustainable development and climate change resilience through the TH model, due to:

- Improvement of the infrastructure by the government (Laisamis-Loiyangalani road funded by AfDB).
- Electrification through a hybrid system by REREC (60 kW) (Rural Electrification Master Plan).
- Construction of a borehole and installation of a pumping system for human and agriculture use by the county government (WARMA access to clean water program).
- Installation of a repeater for the network access by Safaricom (private company intervention).

Laisamis

Laisamis is the main town of Laisamis constituency. It is situated between Merile and Log-logo, on the road from Isiolo to Marsabit Town. In Laisamis most of the buildings are made out of bricks, but most of the people in the surrounding areas are living in huts. Along the road there is the presence of schools, hospitals, shops, hotels, and other business which are benefiting from affordable and reliable energy access.

Laisamis is served by a hybrid mini-grid commissioned by REREC in 2015 that currently serves 170 connected customers, mostly concentrated around the tarmac road and near the town center, where the majority of shops and activities are located. The system that combines 264 kW of diesel generators, 80 kW of solar photovoltaic, and a battery storage system experienced a significant growth of connections thanks to the implementation of the last-mile project. Furthermore, an expansion of the system to reach the communities of Log-logo and Merille started 2 years ago, and the main distribution infrastructure, transformers, and low voltage lines have been put in place in early 2020, but are not yet operational.

Merille and Log-Logo are two urban centers located, respectively, 23 km south and 47 km North of Laisamis. In both towns there is presence of numerous boreholes that are powered by PV systems or diesel generators for shops, dispensaries, and schools. The interconnection of the systems foresees the installation of an additional 500 kW diesel generator to provide adequate electricity supply. This is creating a big opportunity for people to benefit from energy access through income-generating activities.

In Laisamis, access to water is granted by a seasonal river and different private and public boreholes scattered around town. In fact, the water table is not deep, and as results of surveys by the county government and WARMA suggest there is a good potential to dig boreholes. During dry season the sources of water, for the majority of the people, are two public boreholes in the north west of the main settlement, where they can access water for free, but in limited quantities. The salinity level of the water extracted is high and has to be treated accordingly.

Discussion

The selected case studies in Marsabit are evaluated with respect to the theoretical framework laid out in section "The Triple Helix," and in consideration of the literature review detailed in section "Energy Access and Climate Change in Kenya." Although the analysis has been carried out for Marsabit County, a great part of it can be replicated in the other 13 underserved counties identified by KOSAP.

An interesting aspect of this ecosystem is that NGOs have been attempting to improve local conditions by capacity building which can be done by locally present technical schools, vocational training centers, and Technical and Vocational Education Training (TVET) centers. In the proposed cases, in fact, NGOs have developed projects on agriculture and drought resilience species. It was also noted that some electrification activities have also been directly promoted by NGOs, but some projects have failed because the system has not been maintained. Here is when one sees the difference between NGOs and private sector partners. The latter will not leave as long as there is a market for the goods or services rendered.

In the three case studies reported above, water pumping from boreholes or close water sources is significant priority and is usually performed when the village is electrified and water is available nearby. The devolution that started taking place in 2013 has actually helped the county in addressing these problems.

The positive example of Laisamis has shown that combining energy with a safe source of water for people and animals has reduced the nomadism of the area, thus making the electric system more sustainable. Another useful information is an accurate estimate of the population able and willing to purchase energy from the mini-grid. As a matter of fact, load assessment is one of the most critical challenges in mini-grid development.

From the case study analysis, it appears that there is a growing attention in policymaking to the transversally and interdependence of key rural challenges such as energy access, climate change adaptation, and water and food security. Even if not explicitly mentioned, partial references to the TH approach are each day more present in the planning phase though the "silo mentality" that persists.

Education plays a significant role in terms of both enhancing awareness of technologies, opportunities, safe practices, and hygiene in the area and developing skills to promote growth. Currently there are eight vocational training centers (VTCs) in the county that provide above-basic education, four of them instituted by the county government. These centers nonetheless are not yet accredited by the Technical and Vocational Education and Training Authority (TVETA).

It is worth noticing the case of Illaut in which a repeater has been installed shortly after the mini-grid was installed. In fact, people look for connection as a modern mini-grid operator needs connectivity to operate, monitor, and manage the smart-metering system.

Considering this, universities can support and study all the above and identify best practices as well promote new solution to tackle the issues. For example, in Laisamis, the electricity service suffers from poor quality and outages, and the system will be soon enlarged to electrify the nearby villages. However, this may lead to increased number of outages and reduced quality given the long lines to reach the communities. This could be an opportunity for a university research center to provide support by analyzing the data in the field and promoting innovation also by field experience (mode 2). Social data combined with electrical data from the villages can be analyzed in order to estimate and promote the major determinants for load growth and even support the expansion of the mini-grid (Lee et al. 2014).

From the literature review on the TH, it is evident that when government, university, and industry work together in good coordination, rural electrification and climate change adaptation problems are easier to solve.

In a society in which innovation is one of the major drivers for growth, a good application of TH to rural electrification can be not only a promising solution but also promote growth by developing innovation in a challenging and new sector in which few players operate. Besides, electricity provided by decentralized solutions such as mini-grids can be not only an economic solution for areas but provide services that can help people to stop the nomadic lifestyle. Many Turkana people are building permanent houses where the wives and children can stay in a place and profit from education, health service, and electricity while their husbands move around in search of water and grazing fields for their animals.

Rural electrification is a very risky business and rarely can achieve bankability without external funds. However, it brings significant benefits to society in terms of externalities such as local growth, the development of business activities, higher education for children, improved healthcare conditions, and connectivity, to quote but a few services. Therefore, the public sector that looks at society as a whole should be interested in fostering activities and reduce barriers to rural electrification, with tools such as tax exemption, green line of credit, and grants by developing agencies.

The involvement of university in the process can promote innovation by pre-existing knowledge (mode 1), exploiting the field experience (mode 2), and also promoting capacity building for engineers, technicians, and students to provide the desired skills at county level. Therefore, alongside the "mode 2," which justifies the recourse of case studies, the "mode 3" of knowledge production that is based on pluralism and diversity ought to be used.

Additionally, it turns out that the TH concept for rural electrification and climate change might be extended including civil society and concern for natural environment conservation, which have been included respectively as the fourth and fifth helices, respectively, in the work described in Carayannis et al. (2019). Also the concept of "bottom-up triple helix" (Haas et al. 2016) is a valuable contribution toward a better understanding of the role of connector agents, financing institutions, market facilitators, and knowledge brokers. In short, a TH intervention for the development of vulnerable areas promotes the joint action of several entities.

In fact, there is for local universities not only the need to improve their research standards and improvement of curriculums to go beyond the training of workforce but also the mandate to carry out the capacity building of technicians and governmental actors (Perez-Arriaga et al. 2018) and to act as knowledge brokers in order to reach the final rural beneficiaries. In this sense, if we consider that, for example, Nairobi and Marsabit town are located more than 500 km apart, connector agents are needed even within a country in order to promote innovation by experience (mode 2). This specific issue has been addressed in Da Silva et al. (2018) that describes an outreach program to train solar PV technicians on the standards of the Kenyan Energy Regulatory Commission (ERC). The project represents a multi-stakeholder interaction involving academia (Strathmore Energy Research Centre (SERC)), donor agencies (USAID, GIZ), regulatory institutions (National Industrial Training Authority (NITA), ERC) and technical training institutions (TTI) which through "training of trainers" received the capacitation from SERC to train technicians in locations distant from Nairobi.

Conclusions

This chapter has shown that the collaboration of government, universities, and industry in a Triple Helix (TH) framework can be a promising tool for tackling rural electrification and climate change adaptation, using as example the Marsabit County, Kenya, case study.

One such lesson was the need to add to the traditional TH concept the civil society which can provide a soft landing for the private sector partner which at the initial stage may not see significant opportunities for business sustainability.

In modern knowledge-based society, the role of academia in promoting innovation is crucial, and rural electrification sector is not an exception. The university can identify best practices, develop research studies on the possible outcomes of different actions, and even support the development of novel devices, by working closely with the two other partners and having enough data to work with, which are very precious for the effectiveness of the studies.

In Marsabit, which is one of the least developed counties in Kenya, there have been limited examples of TH interventions, but the potential can be significant. Government can provide an adequate policy framework and/or fiscal stimulus, driven by the industry's needs, lowering barriers and reducing risks, after being discussed and agreed with the other stakeholders. Industry can make use of the innovation created by this environment and thrive, hence employing local people and boosting the economy.

The concepts discussed in this study can be useful for policy makers and players in the rural electrification, since they can be easily generalized for different areas in Kenya and Sub-Saharan Africa.

References

African Development Bank Group, Sustainable Energy For All – Africa Hub, Sustainable Energy Fund for Africa (2016) Green mini-grids in sub-Saharan Africa: analysis of barriers to growth and the potential role of the African Development Bank in supporting the sector. GMG MDP Doc Ser n 1

Amaro N, Ruiz C, Fuentes JL et al (2014) Strategic contributions to extreme climate change: the innovation helixes as a link among the short, medium and long-terms. Clim Chang Manag:107–122. https://doi.org/10.1007/978-3-319-04489-7_8

Ambition to Action (2019) The role of renewable energy mini-grids in Kenya's electricity sector. Evidence of a cost-competitive option for rural. electrification and sustainable development

Bagley C, Brown E, Campbell B, et al (2018) Mapping the UK Research & Innovation Landscape: Energy & Development

Blodgett C, Dauenhauer P, Louie H, Kickham L (2017) Accuracy of energy-use surveys in predicting rural mini-grid user consumption. Energy Sustain Dev 41:88–105. https://doi.org/10.1016/j.esd.2017.08.002

Bündnis Entwicklung Hilft (Alliance Development Works), United Nations University – Institute for Environment and Human Security (UNU-EHS) (2016) World Risk Report 2016

Carayannis EG, Acikdilli G, Ziemnowicz C (2019) Creative destruction in international trade: insights from the quadruple and quintuple innovation Helix models. J Knowl Econ. https://doi.org/10.1007/s13132-019-00599-z

Carayannis EG, Campbell DFJ (2010) Triple helix, quadruple helix and quintuple helix and how do knowledge, innovation and the environment relate to each other? A proposed framework for a trans-disciplinary analysis of sustainable development and social ecology. Int J Soc Ecol Sustain Dev 1:41–69. https://doi.org/10.4018/jsesd.2010010105

Da Silva I, Geoffrey R, Nalubega T, Njogu M (2018) Replacing fossil fuel with PV systems through technical capacity building in Kenya. In: Leal Filho W, Surroop D (eds) The Nexus: energy, environment and climate change, green energy and technology. Springer, Cham, pp 171–180

Da Silva I, Wassler S (2011) Implementation of triple Helix clusters procedure in the sub-Sahara Africa energy sector. Case study: academia – CREEC photovoltaic laboratory. In: Conference MPDES 2011

Dzisah J (2011) Mobilizing for development: putting the triple helix into action in Ghana. In: Saad M, Zawdie G (eds) Theory and practice of the triple Helix system in developing countries: issues and challenges. Routledge, New York, pp 146–160

Energies Development (2018) Call for proposals for development of solar PV hybrid mini-grids in Marsabit county

Etzkowitz H (2008) The triple Helix: university-industry-government innovation in action. Routledge, New York

Etzkowitz H (2003) Innovation in innovation: the triple Helix of university – industry – government relations. Soc Sci Inf 42:293–337. https://doi.org/10.1023/A:1026276308287

Etzkowitz H (2012) Silicon Valley: the sustainability of an innovative region. Soc Sci Inf 52:1–25. https://doi.org/10.1177/0539018413501946

Etzkowitz H, Zhou C (2006) Triple Helix twins: innovation and sustainability. Sci Public Policy 33:77–83. https://doi.org/10.3152/147154306781779154

FAO (2017) Resilience analysis in Isiolo, Marsabit and Meru. Anal Resil Target Action FAO Resil Anal 9

FAO (2019) Food security and nutrition in the world

Franz M, Peterschmidt N, Rohrer M, Kondev B (2014) Mini-grid policy toolkit

Fraval S, Hammond J, Bogard JR et al (2019) Food access deficiencies in sub-saharan Africa: prevalence and implications for agricultural interventions. Front Sustain Food Syst 3. https://doi.org/10.3389/fsufs.2019.00104

Gambino V, Del Citto R, Cherubini P et al (2019) Methodology for the energy need assessment to effectively design and deploy mini-grids for rural electrification. Energies 12(12):574. https://doi.org/10.3390/EN12030574

Gibbons M, Limoges C, Nowotny H et al (1994) The new production of knowledge. The dynamics of science and research in contemporary society. SAGE, London

Global Water Partnership Easter Africa (GWPEA) (2015) Assessment of drought resilience frameworks in the horn of Africa. Integr Drought Manag Progr Horn Africa (IDMP HOA

Goldemberg J, Martinez-Gomez J, Sagar A, Smith KR (2018) Household air pollution, health, and climate change: cleaning the air. Environ Res Lett 13. https://doi.org/10.1088/1748-9326/aaa49d

Government of Kenya (2016) Kenya National Adaptation Plan: 2015–2030

Haas R, Meixner O, Petz M (2016) Enabling community-powered co-innovation by connecting rural stakeholders with global knowledge brokers: a case study from Nepal. Br Food J 118:1350–1369. https://doi.org/10.1108/BFJ-10-2015-0398

IEA, IRENA, UNSD, et al (2019) Tracking SDG7: the energy Progress report

International Energy Agency (IEA) (2019a) World energy outlook

International Energy Agency (IEA) (2019b) Africa energy outlook 2019

Kenya Food Security Steering Group (KFSSG), Marsabit County Steering Group (2017) Marsabit County 2017 long rains food security assessment report

Kenya Ministry of Energy (2018) Kenya National Electrification Strategy: key highlights 2018

Kenya National Bureau of Statistics (2019) Kenya Population and Housing Census. https://www.knbs.or.ke/?p=5621. Accessed 1 Mar 2020

KETRACO (2018) KETRACO Grounbreaks Community Projects in Marsabit

Kimatu JN (2015) Evolution of strategic interactions from the triple to quad helix innovation models for sustainable development in the era of globalization. J Innov Entrep 5:0–6. https://doi.org/10.1186/s13731-016-0044-x

Kottász E, Draeck M (2017) Renewable energy-based mini-grids: the Unido experience

KPLC, REA (2017) Kenya off-grid solar access project (KOSAP). Vulnerable & Marginalized Groups Framework

Lake Turkana Wind Power Limited (2011) Addendum on Road Re-Alignment Environmental and Social Impact Assessment (ESIA) Report for the proposed Strengthening of Laisamis-South (D371) and South Horr- Loiyangalani (C77) Road

Lee K, Brewer E, Christiano C et al (2014) Barriers to electrification for "under grid" households in rural Kenya. Center for Effective. Global Action, Berkeley

Leydesdorff L, Etzkowitz H (1995) The triple Helix – university-industry-government relations: a laboratory for knowledge base economics development. EASST Rev 14:14–19

LTWP (2018) Lake Turkana wind power connected to the National Grid

Marsabit County Government (2018) Climate change mainstreaming guidelines: water and sanitation sector

Micangeli A, Del Citto R, Kiva I et al (2017) Energy production analysis and optimization of minigrid in remote areas: the case study of Habaswein, Kenya. Energies 10:2041. https://doi.org/10.3390/en10122041

Ministry of Energy (2018) The Kenya off-grid solar access project. K-OSAP

Ministry of Energy (2019a) KOSAP List of 151 Minigrids – Status after Field Survey (2)- Final (3)

Ministry of Energy (2019b) KOSAP Public Facilities by County - Status Final after Survey (3)

Ministry of Planning and National Development K (2007) Kenya vision 2030: a competitive and prosperous Kenya

Moore A, Nyangoma V, Du Toit J et al (2015) Rwandan collaborative model for educator capacity building. ICSIT 2018 – 9th Int Conf Soc Inf Technol Proc 16:167–172

Murphy B, Corbyn D (2013) Energy and adaptation. Exploring how energy access can enable climate change adaptation. Pract Action Consult

Ngigi S, Busolo D (2019) Devolution in Kenya: the good, the bad and the ugly. Public Policy Adm Res 9:9–21. https://doi.org/10.7176/PPAR

Nygaard I, Gichungi H, Hebo Larsen T, et al (2018) Market for integration of smaller wind turbines in mini-grids in Kenya

Outamha R, Belhcen L (2020) What do we know about university-industry linkages in Africa? In: Abu-tair A, Abdelmounaim L, Khalid AM, Bassam A-H (eds) Proceedings of the II international triple Helix summit. Lecture notes in Civil Engineering Volume, vol 43. Springer, Cham, pp 375–391

Perez-Arriaga I, Micangeli A, Sisul M (2018) Unleashing sustainable human capital through innovative capacity the micro-Grid's potential the micro-grid academy can unlock building and vocational training: in East Africa. Unlocking value from. Sustain Renew Energy:100–105

Republic of Kenya (2019) The energy act, 2019. Kenya Gaz Suppl

Republic of Kenya – The National Treasury and Planning (2018) Third medium term plan 2018–2022. Transforming lives: advancing socio-economic development through the "big four"

Republic of Kenya County government of Marsabit (2018), Second County integrated development plan 2018–2022

Republic of Kenya, County Government of Marsabit (2015) Revised First County Integrated Development Plan (2013-2017)

REREC (2019) Successor of rural electrification authority is now established. Rural Electrification Authority is now REREC

Res4Africa (2019) Water-Energy-Food Nexus – RES4Africa Foundation. https://www.res4africa.org/water-energy-food-nexus/. Accessed 21 Oct 2019

Rural Electrification Authority (REA) (2008) Strategic plan 2008–2012

Smith KR, Bruce N, Balakrishnan K et al (2014) Millions dead: how do we know and what does it mean? Methods used in the comparative risk assessment of household air pollution. Annu Rev Public Health 35:185–206. https://doi.org/10.1146/annurev-publhealth-032013-182356

The Ministry of Agriculture Livestock and Fisheries (MoALF) (2017) Climate risk profile for Marsabit County. Kenya County climate risk profile series. Nairobi

Trenberth KE, Dai A, van der SG et al (2014) Global warming and changes in drought. Nat Clim Chang 4:17–22

Trencher G, Yarime M, McCormick KB et al (2014) Beyond the third mission: exploring the emerging university function of co-creation for sustainability. Sci Public Policy 41:151–179. https://doi.org/10.1093/scipol/sct044

UNICEF (2017) Situation analysis of children and women in KENYA. 224

UNICEF, WHO (2019) Progress on drinking water, sanitation and hygiene. Who:1–66. https://doi.org/10.1111/tmi.12329

Wiesmann U, Kiteme B, Mwangi Z (2016) Socio-Economic Atlas of Kenya: Depicting the National Population Census by County and Sub-Location

World Bank (2020) Kenya Data. https://data.worldbank.org/country/kenya. Accessed 28 Mar 2020

Zhang J, Deng S, Shen F et al (2011) Modeling the relationship between energy consumption and economy development in China. Energy 36:4227–4234. https://doi.org/10.1016/j.energy.2011.04.021

Climate Change Adaptation through Sustainable Water Resources Management in Kenya: Challenges and Opportunities

Shilpa Muliyil Asokan, Joy Obando, Brian Felix Kwena and Cush Ngonzo Luwesi

Contents

Introduction .
East African Community and Climate Change Impacts
Kenya: Country Climate and Water Profile, Climate Change Impacts
Wise Water Management Toward Climate Change Adaptation
Climate Resilience Through Water Management
Water: A Cross-Cutting Factor in Agenda 2030 SDGs
Conclusion
References

Abstract

Water is the medium through which society experiences the most dramatic and direct manifestations of climate change. At the same time, water has a critical

S. M. Asokan (✉)
Climate Change and Sustainable Development, The Nordic Africa Institute, Uppsala, Sweden
e-mail: shilpa.asokan@nai.uu.se

J. Obando
Department of Geography, Kenyatta University, Nairobi, Kenya
e-mail: obandojoy@yahoo.com; obando.joy@ku.ac.ke

B. F. Kwena
Kenya Water for Health Organization, Nairobi, Kenya
e-mail: felix.brian@kwaho.org

C. N. Luwesi
University of Kwango, Kenge, Democratic Republic of Congo
e-mail: cushngonzo@gmail.com

role to play in climate change adaptation and is central towards achieving Africa Water Vision 2025, and the targets set for the 2030 Agenda for Sustainable Development as well as the Kenya Vision 2030. There are fundamental challenges that need to be addressed in order to achieve sustainable water resources management, mainly, the inherent uncertainty associated with the changing climate, the inflexibility in infrastructure and institutions that manage water, and the poor integration of all stakeholders and sectors in water resources management. This study investigates the challenges and opportunities in implementing integrated water resources management and its critical role towards climate change adaptation. A preliminary assessment of sustainable management of water resources and its role in effective climate change adaptation and resilience building in Kenya is carried out through questionnaire survey and stakeholder interactions. Climate change-induced uncertainty, diminishing water sources aggravated by growing water demand, weak institutional and financial governance, and lack of transparency and stakeholder inclusiveness are identified as the main challenging factors that need to be addressed to build a climate resilient society. The study furthermore emphasizes the critical role of water management in achieving Agenda 2030, the Paris Agreement on Climate Change, and the Sendai Framework for Disaster Risk Reduction.

Keywords

Climate change · Water resources management · Climate change adaptation · Climate resilience · Sustainable development · Agenda 2030

Introduction

Our climate is changing and is affecting societies and their livelihoods (IPCC 2018). Climate change translates to water crisis in our everyday lives through the increasing uncertainty in water availability, decreasing water quality, and increasing competition over diminishing water resources (UN-Water 2019). According to the World Economic Forum's 2015 assessment of global risks, water crisis was ranked number one with the potential to severely affect poor and vulnerable population across countries and sectors (WEF 2015). As climate change impacts intensify and the populations and their water demands grow, water stress is likely to increase.

The International Food Policy Research Institute has warned the global community that if business-as-usual continues, by 2050, about 4.8 billion people and approximately half of the global grain production will be at risk because of water stress (Water Resources Group 2016). According to FAO (2013), the rate of increase in water use is more than twice the rate of increase in population during the last century. This increase in water use combined with climate change-induced uncertainty in water availability will aggravate the situation of regions that are already under water stress.

This decade marked a surge in global floods and extreme rainfall events by more than 50% with an occurrence rate that is four times higher than in 1980 (EASAC 2018). According to the Organization for Economic Cooperation and Development (OECD) Environmental Outlook (OECD 2012), the number of people at risk from floods by 2050 is estimated to be about 1.6 billion, and the associated economic risk is expected to be around US$ 45 trillion.

The UN World Water Development Report (WWAP 2019) findings show that the number of people affected and the number of people killed by water-related emergencies such as inadequate water and sanitation, droughts, and flooding are much higher than that due to earthquakes and epidemics and conflicts globally. According to EM-DAT (2019), during the past two decades, floods and droughts have caused more than 166,000 deaths leading to a total economic damage of about US$700 billion.

In Africa, the increase in floods, droughts, and storms has severely affected the livelihoods of population through lack of safe and clean drinking water, crop failures, food shortages, lack of clean energy, etc. According to the World Bank report (2019), Africa has experienced increasing climate and water risks with more than 2,000 natural disasters since 1970, of which about 1,000 happened in the last decade. According to OECD (2012), between 75 and 250 million people in Africa are projected to be under increased water stress, and agricultural production can reduce by up to 50% in some regions by 2020. The rainfall projections in Africa indicated that Southern Africa will become drier and Eastern and Western Africa will become wetter with increased risk of floods (UNDP 2018). The recent water crisis in South Africa, the food crisis in Sahel, and disastrous cyclones in Mozambique are reminders of society's vulnerability and unpreparedness in the wake of climate change reality, which in most of the cases are aggravated by incompetent institutional and financial capacities.

According to the UN Environment Adaptation Gap Finance Report (UNEP 2018), the global estimates of the costs of climate change adaptation in developing countries are between US$140 billion and US$300 billion by 2030 (which is two to three times higher than the earlier estimates cited in the Intergovernmental Panel on Climate Change [IPCC] report). By 2050, the estimates can plausibly be four to five times higher – US$280 billion to US$500 billion. For Africa, this would mean that the resources that would have otherwise directed for economic growth, to overcome poverty and to achieve the United Nations Sustainable Development Goals [SDGs] are now to be diverted toward climate change adaptation and resilience building initiatives.

East African Community and Climate Change Impacts

Climate model projections for East Africa indicate that rainfall and temperature will become extreme and intense in the future affecting both the quantity and quality of surface and groundwater in the region. This will severely affect the accessibility to safe and clean drinking water, health, food security, and energy resources (EAC and USAID 2018; Luwesi and Di Luyundi 2019). Adapting to these changes will require not only accurate knowledge of the frequency and severity of the extreme events but also planning and implementation of critical interventions to curb disasters across the East African Community (EAC). Moreover, this is highly important for the smooth implementation of the water targets of the African Union Agenda 2063, the Africa Water Vision 2025, and the 2030 Agenda for Sustainable Development, including the Kenya Vision 2030. The EAC has thus a long way to go for eliminating poverty, hunger, and food insecurity to give its citizens the right to healthy and productive lives.

Kenya: Country Climate and Water Profile, Climate Change Impacts

The National Water Policy of Kenya was developed in 1999 and based on that Water Act was established in 2002 (National Water Master Plan 2030 2013). The constitutional reforms of 2010 in Kenya introduced decentralization policy which came into effect in 2013 (Alexis and Lumbasi 2016). Since then the country has undergone significant transition through devolving governance to the 47 newly established counties. Moreover, the country's 2016 Climate Change Act formulated a comprehensive law and policy to guide national and subnational responses to climate change. The Water Resources Management Authority (WRMA) was reformed to Water Resources Authority (WRA) with a mandate that includes regulation, protection, and dissemination of information on water resources. Furthermore, WRA undertakes climate actions in terms of mitigation and adaptation to minimize the effects of global warming and climate change.

With the projected increase in temperature of about 2.5 °C between 2000 and 2050, and with rainfall becoming more intense and less predictable, Kenya is considered as very vulnerable to the impacts of climate change. The arid and semiarid lands (ASALs) in the north and the east of the country are particularly at risk because of the increased occurrence of drought and the associated food and water security challenges. The increased frequency of droughts, floods, and landslides in the Rift Valley Province and the increased floods and saltwater intrusion in the coastal areas are also posing challenges that demands efficient actions at national level to improve the adaptive capacity and resilience of the population (Government of the Netherlands 2019).

About 80% of the country's geographical area is arid or semiarid, and the main source of sustenance is pastoral and subsistence agriculture. The livelihoods of the poor are severely affected by the climate crisis and are having a severe impact on country's socioeconomic development (Mwendwa and Giliba 2012). A questionnaire survey conducted by Mwendwa and Giliba (2012) found recurrent droughts and changes in rainfall patterns to be the most important indicators of the effects of climate change and food shortage, hunger, famine, water scarcity, low yields, and high poverty levels as the most important impacts of climate change in Kenya.

According to Ng'ang'a (2006), the 2004/2005 drought resulted in a prolonged famine in the less rainfall areas in the country. Decreasing lake surface area, for example, in the case of Lake Naivasha, has affected water quality, quantity, and navigability and has affected its biodiversity (Mironga 2005).

In urban areas, the impacts of climate change are visible mainly in informal settlements, where the risks arising from droughts and floods are severe. The risks in rural areas arise because of the dependence of rural population on the increasingly diminishing and degraded water sources. The major challenges faced by the peri-urban and rural population are the limited management capacity and low operating revenues. According to Mureithi et al. (2018), the lack of access to finance is also a major constraint as funding tends to be generally allocated to develop new water systems in very low-income areas with poor access rather than on restoring an already existing but poorly performing water supply scheme.

Wise Water Management Toward Climate Change Adaptation

It is crucial to acknowledge that water is not just a sector that has been impacted by climate change. The central role of water in adapting to the climate change-driven variations has been increasingly recognized over the past decade. However, water's ability to address climate change adaptation still faces many challenges. The inherent uncertainty associated with the changing climate, the inflexibility in infrastructure, and institutions that interact with water, in addition to the lack of an integrated approach in water resources management, are the fundamental challenges that need to be addressed at global, regional, and local scale (UN-Water 2019).

Water is inextricably linked with food and energy. Addressing climate change adaptation through effective water management is an opportunity by which the governance, infrastructure, and financing mechanisms can be transformed, leading to a holistic development of the community. It is highly important that the national policies should reflect on climate change adaptation strategies that are being followed in similar hydroclimatic parts of the world and at the same time incorporate the locally developed adaptation techniques to improve its climate resilience. Increased water stress coupled with increasing future water demands will require tough decisions on how to allocate water resources between competing water uses, including for climate change mitigation and adaptation.

Wise water management can be achieved through effective, efficient, and partic-ipatory management of water resources. Scarcity of water leading to drought can be addressed through increased rain water harvesting and water storage to capture the minimal available water at the source. These water harvesting structures will also aid in enhancing groundwater recharge, thereby improving the resilience of the society to future low surface water availability. In order to address the problems of excess water because of floods, the Water Resources Authority is in the process of devel-oping plans to prepare the society towards flood disasters.

Application of innovative climate technology on the ground through sensor data, models, and weather forecasting improved water distribution systems, avoiding leakages and thereby decreasing loss through nonrevenue water are delivering positive outcomes. According to a recent report (Global Center on Adaptation 2019), researchers at Jomo Kenyatta University of Agriculture and Technology in Kenya in collaboration with Wageningen University and Research in the Netherlands is planning to launch the Climate Atlas, a localized weather monitoring system, with the aim of providing local rainfall and temperature projections across the 47 counties in Kenya for the period 2050–2100. The overall aim of this project is to protect Kenya's food supply for the future years. Another report by UNFCCC (2019) discussed a readiness proposal by Ghana on "Integrated Climate Monitoring and Early Warning System" aimed at strengthening the country's capacity to build an early warning system for droughts. In cities, nature-based solutions through restoration of degraded ecosystems in the buildup areas can improve the adaptive capacity. The city area can be envisaged as a catchment unit, and the effective capturing and managing of stormwater can make the city more resilient.

Climate Resilience Through Water Management

Adapting to climate change after the occurrence of a disaster is a reactive approach. Building resilience on the other hand is the ability to adapt to and transform and recover from that hazard in a timely, efficient, and sustainable manner. The first step towards achieving resilience is to have access to the relevant information, for example, through public campaign or by collaborating with established networks. Participatory vulnerability assessments at grassroots level are critical. Furthermore, resources should be available to make assessment and to ensure and facilitate processes leading to resilience. For instance, one need to prioritize the most critical climate change impact in a region based on the exposure, sensitivity, and adaptive capacity of the population to that impact. Based on that, a monitoring and evaluation system has to be designed and organizational capacity need to be build. Ensuring political commitment as well as facilitating implementation of public private partnerships can aid in short-term development of resilience. The final step is to develop necessary institutional arrangements that support long-term adaptation measures.

In Kenya, the relevant policy and legislative framework has been put in place to guide the country's response to the challenges of climate change. Climate change adaptation and resilience has been set in the National Adaptation Program (NAP) and is operationalized through the National Climate Change Action Plan (NCCAP). This is done by mainstreaming adaptation across all sectors in the national planning, budgeting and implementation processes, and taking cognizant of the fact that climate change is a cross-cutting sustainable development issue with economic, social, and environmental impacts (GoK 2016). The other policy documents include the National Drought Management Authority Act 2016, National Policy for Disaster Risk Management 2013, the National Climate Change Response Strategy (NCCRS) 2010, and the Climate Change Act 2016.

Climate resilience through effective water management can be achieved through sustainable implementation of integrated water resources management. Functional water governance mechanisms can reduce disasters and build capacity for adaptation and thereby improve resilience. Locally designed and implemented water storage and flood control structures and nature-based solutions, for instance, lakes as natural water storage structures and floodplains as natural excess runoff absorbers, are some examples on climate resilience through wise water management. Healthy ecosystem services are dependent on well-functioning river basins, which in turn can support agriculture and fisheries, wastewater treatment, drinking water provision, groundwater recharge, and coastal protection among others. In Zambia, the natural underground water reservoirs were protected through proper waste disposal, hence improving aquifer health and usage, thereby benefiting the local population in Lusaka (IWA 2019).

Building resilience through empowerment of the climate change-affected community has been reported in many parts of the world. One such example is the Osukuru United Women's Network in Uganda (Global Resilience Partnership 2019). The major climate change impact faced by this community was flooding and the associated displacements and health issues. This women's network initiated campaigns and improved the resilience of the community by building trenches around the houses to prevent floodwater from entering their homes. With more and more people joining this network, the success rate is high. However, there are challenges in the sustainable functioning of the network, for instance, financial viability, capacity building, training, mobility, etc. Enhanced climate resilience thus includes strong economic growth, resilient ecosystems, and sustainable livelihoods. It is noted that through climate financing (Odhengo et al. 2019) and public-private partnerships, there will be opportunities for progress in Kenya

The core approach in building resilience includes creating awareness, building social capital, improving technical capacities, and thereby empowering the community. Community-based vulnerability assessment and development of resilience action plan and promoting joint action through multi-stakeholder groups aid in faster adaptation. Furthermore, by making the livelihood-dependent economies diverse and flexible can also improve the community's resilience to climate change. Knowledge, learning, innovation, and clean technology are key towards achieving climate resilience. Smart agricultural technology such as integrated weather and market advisories inform farmers on what and when to grow and harvest, where to sell their produce, etc.

Questionnaire survey and stakeholder discussions including governmental, nongovernmental, and other organizations in Kenya are ongoing. According to the survey results, climate change impacts that are being experienced in the region are manifold – floods, droughts, storms, and its associated impacts on the society and the ecosystem. One major impact is on agriculture and food production, especially on the subsistence farming. Decline in agricultural produce has reduced the marginal GDP in the agriculture sector. This along with the fluctuations in world market prices and changes in geographical distribution of trade regimes has caused an increased number of people at risk of hunger and food security. Extreme weather conditions have affected water supply and water quality, thereby affecting human health. The increased demand coupled with the decreasing supply of source water resources has led to conflicts over water sharing. The timely dissemination of relevant climate information to the most vulnerable sections of the society is yet another major challenge that needs to be addressed.

The sustainable management of water resources faces major challenges because of climate change-induced uncertainty, diminishing water resources, increased water scarcity, and lack of protection of water sources through regulation. Community engagement and stakeholder participation especially from the vulnerable communities are crucial, and ownership and control of water resources are found to have resulted in more active participation of stakeholders. Incorporation and mainstreaming of innovation and technology are increasingly applied but have more scope for improvement. Climate change adaptation through rainwater harvesting, water storage development, and climate smart agriculture is followed by communities with support from government. The major constraint in the wise management of water resources is the long-term sustainable finance and funding mechanisms.

Kenya has identified climate change and disaster risk management as two of the three thematic areas in their National Plan. The nationally determined contributions (NDC) identify mainstreaming of climate change adaptation in the water sector as one of the priority adaptation strategies (Ministry of the Environment of the Republic of Kenya 2013). Adaptation strategies include development of water resources monitoring and early warning assessments.

Water: A Cross-Cutting Factor in Agenda 2030 SDGs

The achievement of Agenda 2030 Sustainable Development Goals are dependent on improved and sustainable water management (UNESCO and UN-Water 2020). There are multiple interlinkages of SDG 6 on water to the other 16 SDGs and also intralinkages within the SDG 6 connecting water and sanitation, water quantity and quality, IWRM, and community engagement in IWRM. The institutional and financial investment in achieving SDG 6 through resilient water management will inherently advance the progress of other SDGs through: overall poverty reduction (SDG 1); water security supporting the food and agriculture sector and thereby eradicating hunger (SDG2); quality water toward good health (SDG3); safe, segregated toilets and menstrual hygiene improves girls' access to education (SDG 4); gender equality in public and working life (SDG 5); access to energy and hydropower as one solution for sustainable energy (SDG 7); inclusive and productive economic growth and employment through water (SDG 8); resilient and sustainable water infrastructure (SDG 9); reduced inequalities through accelerated and inclusive action on water supply and sanitation (SDG10); sustainable and equitable urban development (SDG 11); equitable use and efficient water resources management (SDG 12); combating climate change impacts through climate-resilient water resources management (SDG 13); water quality and water resources management upstream on the land and along the rivers impacts coastal and marine ecosystem (SDG 14); increasing sustainable management of soil (SDG 15); promoting peaceful societies and accountable and inclusive institutions through effective water governance (SDG 16); and building commitments and accountability in the water sector, institutional coordination of water programs, mobilizing funding, and improved technology through strengthened means of implementation (SDG 17). Acknowledging the significance of water in these interlinkages and increasing efforts in addressing the challenges to achieve SDG 6 will automatically advance progress in climate change adaptation through wise water resources management and will help build resilient communities.

Although water is not explicitly mentioned in the Paris Agreement or in the Sendai Framework, it is important to acknowledge its relevance in achieving most of the mitigation and adaptation strategies and targets. The United Nations Framework Convention on Climate Change (UNFCCC) assessed the link between climate change and integrated water resources management such as watershed protection, waste- and stormwater management, water conservation, recycling, and desalination (UNFCCC 2019), where water governance is emerging as a leading domain for resilience to climate change.

Conclusion

Climate change primarily impacts society through several disasters, such as floods, droughts, storms, etc. which then severely affects human security and socioeconomic development of the region. Building and enhancing resilience to these disasters through effective water management are crucial, and water can play a major role in achieving effective climate change adaptation. Furthermore, disasters are recognized as opportunities to revitalize livelihoods, environment, and economies that can then lead to communities that are more resilient.

Innovative technologies in water supply and management and active participation of civil society, government, and private sectors can improve urban resilience. An integrated water resources management is imperative for attaining both rural and urban resilience as it safeguards their livelihoods and food security. It is also essential to ensure inclusiveness and meaningful participation of all stakeholders and include indigenous adaptation practices and traditional knowledge.

National and regional climate policy and planning should follow an integrated approach to climate change and water management. It is extremely important that the national adaptation strategies assist the vulnerable sections of the society and improve their resilience to the climate change impacts. Furthermore, increased investment is needed in institutions, capacity development, better data collection, assessment, and sharing.

References

Alexis S, Lumbasi L (2016) The impact of decentralization in Kenya, Trinity College London, Masters in Development Practice 2016, AidLink enabling communities in Africa tackle poverty
EAC, USAID (2018) Climate change vulnerability and adaptation in East Africa, current and future climate change. https://www.climatelinks.org/sites/default/files/asset/document/2018_USAID-

PREPARED-TetraTech_VIA-East-Africa-Current-Future-CC-Factsheet.pdf. Accessed 27 Mar 2019

EASAC (European Academies' Science Advisory Council) (2018) Extreme weather events in Europe: preparing for climate change adaptation: an update on EASAC's 2013 study. https://easac.eu/publications/details/extreme-weather-events-in-europe/

EM-DAT (Emergency Events Database) (2019) The emergency events database. Centre for Research on the Epidemiology of Disasters (CRED), Université catholique de Louvain, Brussels. www.emdat.be

FAO (2013) Coping with water scarcity: an Action framework for agriculture and food security. FAO water reports no. 38. FAO, Rome. www.fao.org/3/a-i3015e.pdf

Global Center on Adaptation (2019) How mapping changes on a local level will help Kentan farmers. https://gca.org/solutions/how-mapping-changes-on-a-local-level-will-help-kenyan-farmers?utm_sq=g9atrsd0w0. Accessed 27 Mar 2020

Global Resilience Partnership (2019) Thriving in the face of surprise, uncertainty and change, Resilience Insights 2019. http://grpinsightsreport.info/. Accessed 31 Oct 2019

GoK (2016) Kenya National Adaptation Plan: 2015–2030, Enhanced climate resilience towards the attainment of Vision 2030 and beyond, Government of Kenya, July 2016.

Government of the Netherlands (2019) Climate change profile, Kenya. Report by the Ministry of Foreign Affairs, Government of the Netherlands.

IPCC (2018) Summary for policymakers. Global warming of 1.5°C. An IPCC special report on the impacts of global warming of 1.5°C above pre-industrial levels and related global greenhouse gas emission pathways, in the context of strengthening the global response to the threat of climate change, sustainable development, and efforts to eradicate poverty. IPCC, Geneva. www.ipcc.ch/sr15/chapter/spm/

IWA (2019) Nature-based solutions: the advantage of being a developing country. https://iwa-network.org/nature-based-solutions-the-advantage-of-being-a-developing-country/. Accessed 30 Oct 2019

Luwesi CN, Di Luyundi TM (2019) A quick appraisal of the impact of environmental changes on undernutrition in Kenge Health Zone, DRC. Biomed J Sci Techn Res 18(3):19–26. https://doi.org/10.26717/BJSTR.2019.18.003145

Ministry of the Environment of the Republic of Kenya (2013) The National Water Master Plan 2030. The Republic of Kenya. https://wasreb.go.ke/national-water-master-plan-2030/

Mironga JM (2005) Environmental implications of water hyacinth infestation in Lake Naivasha, Kenya. In: 11th World lakes conference, Nairobi, Kenya, 31st October to 4th November 2005, Pp: 580 - 592, Proceedings Volume II, Ministry of Water and Irrigation International Lake Environment Committee.

Mureithi PM, Luwesi CN, Mutiso MN, Förch N, Nkpeebo AY (2018) Legal and market requirements for water finance. In: Beyene A, Luwesi, CN (eds) Africa: the Kenya water sector reforms case in innovative water finance in Africa, A guide for water managers, volume 1: water finance innovations in context. The Nordic Africa Institute, Uppsala, Sweden.

Mwendwa P, Giliba RA (2012) Climate change impacts and adaptation strategies in Kenya. Chin J Popul Resour Environ 10(4):22–29. https://doi.org/10.1080/10042857.2012.10685104

National Water Master Plan 2030 (2013) The project on the development of the National Water Master Plan 2030, final report volume 1, Executive summary

Ng'ang'a JK (2006) Climate change impacts, vulnerability and adaptation assessment in East Africa, UNFCCC African regional workshop on adaptation, 21–23 September 2006, Accra

Odhengo P, Atela J, Steele P, Orindi V, Imbali F (2019) Climate finance in Kenya: review and future outlook, discussion paper

OECD (2012) OECD environmental outlook to 2050, OECD Publishing. https://doi.org/10.1787/9789264122246-en. Accessed 30 Oct 2019

UNDP (2018) Climate change adaptation in Africa, UNDP synthesis of experiences and recommendations. UNDP/GEF, New York

UNEP (2018) The adaptation gap report 2018. United Nations Environment Programme (UNEP), Nairobi, Kenya. https://www.unenvironment.org/resources/adaptation-gap-report

UNESCO, UN-Water (2020) United Nations World Water Development report 2020: water and
    climate change. UNESCO, Paris
UNFCCC (2019) How developing countries are scaling up climate technology action, Article 19
    Nov 2019. https://unfccc.int/news/how-developing-countries-are-scaling-up-climate-technol
    ogy-action
UN-Water (2019) Climate change and water, UN-Water policy brief. https://www.unwater.org/
    publications/un-water-policy-brief-on-climate-change-and-water/. Accessed 28 Oct 2019
Water Resources Group (2016) Annual report, Partnerships for transformation, water security
    partnerships for people, growth and the environment
WEF (2015) Global risks 2015, 10th edn. World Economic Forum. http://www3.weforum.org/docs/
    WEF_Global_Risks_2015_Report15.pdf. Accessed 18 Nov 2019
World Bank (2019) This is what it's all about: building resilience and adapting to climate change in
    Africa. https://www.worldbank.org/en/news/feature/2019/03/07/this-is-what-its-all-about-build
    ing-resilience-and-adapting-to-climate-change-in-africa?cid=EXT_WBEmailShare_EXT.
    Accessed 31 Oct 2019
WWAP (2019) The United Nations World Water Development report 2019, Leaving no one behind,
    UNESCO World Water Assessment Programme. UNESCO, Paris

Climate Change Impact on Soil Moisture Variability: Health Effects of Radon Flux Density within Ogbomoso, Nigeria

Olukunle Olaonipekun Oladapo, Leonard Kofitse Amekudzi, Olatunde Micheal Oni, Abraham Adewale Aremu and Marian Amoakowaah Osei

Contents

Introduction
Climate Change and Soil Moisture
Radon and Human Health
Silmulation and Forecasting of Soil Moisture Within Ogbomoso Using the Swat Mode
    The SWAT Model
    Soil Moisture Observatio.
    Soil Moisture Simulation
    Forecasted Soil Moisture
Estimation of Radon Flux Density
Conclusion
References

O. O. Oladapo (✉)
Department of Science Laboratory Technology, Ladoke Akintola University of Technology, Ogbomoso, Nigeria
e-mail: oooladapo66@lautech.edu.ng

L. K. Amekudzi · M. A. Osei
Department of Physics, Kwame Nkrumah University of Science and Technology, Kumasi, Ghana

O. M. Oni
Department of Pure and Applied Physics, Ladoke Akintola University of Technology, Ogbomoso, Nigeria
e-mail: omoni@lautech.edu.ng

A. A. Aremu
Department of Physics with Electronics, Dominion University, Ibadan, Oyo, Nigeria

Abstract

Climate affects the quantity of soil moisture within the surface of the earth and this is obtained by affecting the amount of radon flux density escaping from the land surface. This chapter contains the evaluation of climate change conditions as it affects the variability of soil water for the purpose of estimating the health effects of radon flux density within Ogbomoso metropolis. The simulated soil moisture content around Ogbomoso was done for a period of 34 years using the hydrological model, Soil Water Assessment Tool (SWAT). The calibration and validation of the SWAT model was done using the daily observed soil moisture content. The simulated daily soil moisture within Ogbomoso showed good performance when calibrated and validated. A 20 years prediction of the daily soil moisture content was done using the SWAT model. The estimation of the radon flux density for the study area was obtained using the simulated soil temperature and soil moisture from the SWAT model. In this chapter, the UNSCEAR radon flux formula was used for the radon flux estimate. The result showed that the UNSCEAR radon flux formula performed well in estimating the radon flux density in the study area. The mean value of the radon flux density of $15.09 \text{ mBqm}^{-2} \text{ s}^{-1}$ falls below the estimated world average of $33 \text{ mBqm}^{-2} \text{ s}^{-1}$ by UNSCEAR stipulated for land surface. The results showed that Ogbomoso region is not prone to high risk of radon exposure to the public. The estimation of the radon flux density value suggested that there is no radiological health hazard such as lung cancer or any other respiratory tract diseases to the inhabitant of Ogbomoso, Nigeria.

Keywords

Climate change · Soil moisture · Radon flux density · Estimating · SWAT model · Ogbomoso · Simulating and forecasting

Introduction

Climate change has been reported to have significant effects on the amount of soil moisture within the land surface (Schery et al. 1989). This is attainable by managing the quantity of radon emanating from the earth. The main hydrology and climate variable that affects the land surface processes is soil moisture (Tanner 1980; Schery et al. 1989). Soil moisture is a major variable for a large number of uses, including numerical weather forecasting, flood prediction, agricultural drought evaluation, water resources management, and health evaluation (Zhang et al. 2008). Despite the principal importance of soil moisture to agricultural performance, crop observation, and yield prediction, soil moisture statistics is not widely obtainable on a territorial scale in Nigeria. Only finite data sets of soil moisture exist in Nigeria because it is a strenuous property to measure. Soil moisture is naturally diverse because of dissimilarity in moisture-holding abilities of the soil at a very small scale and topography. The soil moisture is mostly measured by agro-meteorological experimental stations which are very insufficient in Nigeria. The data obtained from these stations are too few to investigate how the atmosphere relates with the land surface because they are measured basically for agriculture. Apart from these, agro-meteorological experimental stations are not available in Ogbomoso metropolis. It is paramount to seek a way to simulate this limited variable.

Investigation has shown that variation in soil moisture have evidential effects on changes in radon concentration (Radon-222) in the atmosphere (Brookins 1990). Radon is a natural radioactive inert gas whose inhalation in excess dose may lead to adverse health effect such as lung cancer. Factors such as climatic parameters, diffusivity, and radon concentration have been identified as some of the factors that determined the concentration of radon in the atmosphere (Zahorowski et al. 2004; Fields 2010). Soil moisture is a major parameter that determines the amount of radon exhalation from the earth surface and radon concentration in the atmosphere (Tanner 1980; Ball et al. 1983; Zhou et al. 2005; Sheffield and Wood 2008). The purpose of this chapter is to assess the health effects of radon flux density due to the impact of climate change on soil moisture within Ogbomoso metropolis, Nigeria. This chapter will provide an all-inclusive, evidence-based, measurable estimation of simulated and forecasted soil moisture for assessing radon flux–related health effect in Ogbomoso, Nigeria.

In this chapter, the simulation and forecasting of the soil moisture from the available meteorological data and other relevant data using the Soil Water Assessment Tool (SWAT) model was investigated to estimate the health effects of radon flux density distribution in the atmosphere. This chapter, therefore, seeks to use the relationship between soil moisture and radon emanation from the ground surface to estimate radon flux density distribution in the atmosphere within Ogbomoso. The objective of this chapter is to investigate the effects of change in climate on the variability of soil moisture for the purpose of estimating radon flux density in the Ogbomoso, Nigeria. The specific objectives of this chapter are to obtain a quantitative estimation of (a) the simulated soil moisture, (b) the forecasted soil moisture, (c) the radon flux density above the land surface, and (d) to investigate the health risk of the estimated radon flux density.

Climate Change and Soil Moisture

In the interaction of land and the atmosphere, the amount of water in the top soil plays a very principal role (McColl et al. 2017; Koster et al. 2004). Soil moisture is propelled by climate, more significantly by precipitation and temperature (Feng and Liu 2015). Precipitation is the major source of soil moisture, and changes in precipitation will impact the amount of water in the top soil. Furthermore, changes in temperature also affect soil moisture by managing evapotranspiration (Wang et al. 2018a). The application of soil moisture for weather prediction (Alexander 2011; Wang et al. 2018b), drought observation (AghaKouchak 2014), hydrological modeling (Wanders et al. 2014), and vegetation variation (Chen et al. 2014) are gaining wide application in recent time. The global warming today and movement of the water cycle may cause changes in soil moisture variability (Sheffield and Wood 2008). It has been reported that changing vegetation can also alter the amount of water in the top soil (Sterling et al. 2013). Vegetation variation has been discovered to affect field ability and soil penetration, thus altering soil moisture (Ouyang et al. 2018). These factors, through some complex relationships, alter soil moisture. Climate variation has been discovered to impact soil moisture greatly. The effects of change in climate and change in vegetation have been reported in literatures.

This chapter considers the implications of change in climate on soil moisture for estimating the health effect of radon flux density. The various soil moisture measuring stations, which are few in number, do not consider the different soil types, soil characteristics, land cover and land use, topography, etc. on the soil moisture. The observation of soil moisture is very scarce. Instead of making use of observation that is limited, numerical models have been employed over time to investigate the changes in hydrological processes and soil moisture, which play a key role in this regard. Numerous hydrological models exist today that can be used for the purpose of simulating the hydrological processes of a watershed. Each model possesses its own pros and cons. The purpose to which an investigation is carried out determines the choice of model to be used. This chapter seeks to examine the Soil Water Assessment Tool (SWAT) model (Arnold et al. 1998) which was chosen due to its ability to simulate hydrologic processes. Another strength of the SWAT model is the fact that it takes into consideration the effects of vegetation properties, topography, soil characteristics, and land type and land cover on the hydrological processes of the soil. This is made possible with the aid of a Digital Elevation Model (DEM) with a very high resolution. Over the years, the SWAT model has become accepted on a global scale as a strong modeling tool suitable for simulating watershed hydrology (Gassman et al. 2007). The SWAT has found application in extensive scope of environmental situations, watershed scales, and framework analysis as described by Gassman et al. (2007). In this chapter, some meteorological data collected from NIMET for Ogbomoso metropolis were fed into the SWAT model to simulate and forecast soil moisture over Ogbomoso. The simulated soil moisture data were used to estimate radon flux density, thereby examining the health effect over Ogbomoso region.

Radon and Human Health

Over the years, radon gas has been studied for two primary purposes. The first purpose is to investigate the extent to which the general public is exposed to it. Secondly, it was studied to detect the various transfer processes in the atmosphere. The breathing in of radon and its momentary progeny account for about half of the effective dose from all natural sources of ionizing radiation. One of the main challenges of radiation protection to the public is the development of lung cancer as a result of inhalation of the progeny of radon (UNSCEAR 1993). Whenever radon is inhaled in high concentration, it usually leads to diverse health implications.

Atmospheric radon, which is an inert gas with a half-life of 3.82 days, has been identified as detectors for atmospheric transport. It also doubles as indicators for evaluating and simulating environmental activities (Brookins 1990; Iida et al. 1996; Wang et al. 2004, Zahorowski et al. 2004; Ohkura et al. 2009). Radon concentration in soil and radon exhalation from the soil surface depend on many physical characteristics. These characteristics are related to soil parameters, such as radium content and the internal structure of the soil. Other physical parameters include type of the mineralization, soil porosity, grain size of the soil, permeability of the soil, and emanation coefficient (Mazur et al. 1999; Aburnurad and Al Tamimi 2001). The release of radon gas from the soil to the atmosphere is associated with some of the same processes that control the soil/air exchange of important greenhouse gases like CH_4, CO_2, and NO_2. Radon does not go through complex chemical reactions and its source term is relatively well known (^{226}Ra in soil).

Measurement of radon concentration in the atmosphere has been used for the validation climate models. Due to insufficient radon flux density measurement, measurement of atmospheric radon has been employed in the validation radon flux density from the soil (Kritz et al. 1998; Gupta et al. 2004; Zhang et al. 2008). Therefore, statistics on the territorial distribution of ^{222}Rn exhalation from the earth's surface is regarded as useful for recognizing areas with a health risk of high radon exposure to public. This chapter therefore aimed to use the simulated and forecasted soil moisture to estimate the health effects of radon flux density of Ogbomoso metropolis. Studies of the radon flux densities are able to provide information on the interaction of exchange of gases between the soil and the atmosphere for the purpose of estimating pure health implications of the inhalation of these gases.

Silmulation and Forecasting of Soil Moisture Within Ogbomoso Using the Swat Model

The SWAT Model

Soil moisture, which is the amount of water in the upper layer of the soil, has been proven to be affected by changes in climatic properties due to alteration in some meteorological parameters such as precipitation, temperature, relative humidity, solar radiation, and wind speed. The alterations in these parameters are employed in the simulation and forecast of the soil moisture over a particular region provided that relevant data for that region are available. The Soil Water Assessment Tool model known as the SWAT model was used for this purpose within Ogbomoso metropolis. Ogbomoso, which is geographically situated within 4°10′E to 4°20′E longitude and 8°00′N to 8°15′N latitude, is investigated in this chapter. The area is situated within the crystal-like Vault of Nigeria (MacDonald and Davies 2000). In Ogbomoso, rocks are classified as either quartzites or gneisses (Ajibade et al. 1988). The agricultural watershed considered in this research work where the simulation and forecasting of soil moisture was carried out is situated within Ogbomoso North Local Government Area as shown in Fig. 1 (Adabanija et al. 2014). Simulation of

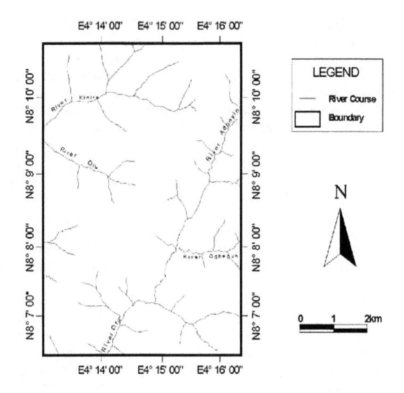

Fig. 1 The map showing the area within the agricultural watershed (Adabanija et al. 2014)

hydrology within this agricultural watershed was considered under diverse topography, vegetation, climatic condition, and soil. The total drainage of the Ogbomoso watershed accounts for a total of 8.58 km². Ogbomoso, which lies between the savannah and the rain forest, has wet and dry tropical climate. The temperature of the region is fairly uniform throughout the year. It has an average rainfall and temperature of 1200 mm and 26.20 °C annually, respectively. The territorial district of Ogbomoso has four seasons similar to other southern Nigeria. The initial raining season in the region starts from April to July characterized by high humidity and heavy rainfall. It is usually interrupted by a short dry period in August before the rainy season resumes again, this time for a short wet season between September and October. The fourth season is the harmattan season which spans between November and ending in the middle of March. The region experiences a relative humidity between 75% and 95%. Ogbomoso region, being a low land forest, has agricultural activities as major activities of the inhabitants. Figure 1 shows the region of investigation within Ogbomoso metropolis.

"The SWAT has been employed to simulate watersheds globally, it is most frequently chosen due to its vigorous ability to compute the impacts of land management practices on hydrological processes, water quality, and crop growth" (Arnold et al. 1998). In this chapter, the QSWAT, which is a type of the SWAT2012 model, assembled with QGIS 1.4 was employed. The SWAT model operates by breaking a watershed into sub-basins, connected by a stream grid. The Hydrological

Table 1 Input database and sources for the SWAT model

Data Type	Source
Digital Elevation Model	SRTM 1-Arc-Second Global v3 (30 m)
Land-Use Data	MODIS (15-arc) (Broxton et al. 2014)
Soil Data	Digital global soil map FAOv3.6
Meteorological data	Nigerian Meteorological Agency (1984–2017)
Climate Projection Data	Canadian Regional Climate Model (2018–2037)

Table 2 Statistical model performance for the calibration and validation procedure (Oladapo et al. 2018)

Procedure	Regression coefficient (R^2)	Nash and Sutcliffe efficiency (E_{NS})	Percentage difference (D)
Calibration	0.91	0.64	13
Validation with ESA CCI	0.81	0.53	11
Validation with in situ measurement	0.88	0.84	8
Standard	>0.6	>0.5	≤15%

Response Unit (HRU) is the place where the simulation was carried out, which is logged in each basin. The simulated variables (water, sediment, nutrients, and other pollutants) are first simulated at the HRU and then passed through the stream to the water outlet (Arnold et al. 1998; Neitsch et al. 2005). The study on soil moisture simulation and forecasting was first investigated within Ogbomoso by Oladapo et al. (2018) with the aid of climatic data inputs such as precipitation, maximum and minimum temperature, solar radiation, relative humidity, and wind speed. Secondary data of daily precipitation, maximum temperature and minimum temperature data, solar radiation, the humility, and the wind speed were all collected from the Nigerian Meteorological Agency. The data cover between 1979 and 2017. Other data sources including the digital elevation model, land use data, soil data, and climate progression data of the study were used in the simulation and the forecast of the soil moisture in the area (Table 1).

Since Ogbomoso is a natural hydrological setting, five years daily satellite-based soil moisture data (2006–2010) from ESA CCI soil moisture project for the region under investigation were used for the SWAT model calibration. The laid down method of calibration for the SWAT model was carefully followed (Santhi et al. 2001; Neitsch et al. 2005). Due to lack of long-term in situ measurement of soil moisture at the study area, both the short-term in situ measurement of daily soil moisture taken at the study area between April and July 2017 (Oladapo et al. 2018) and five years daily soil moisture data from ESA CCI (2011–2015) were used to validate the SWAT model (Oladapo et al. 2018). Three different model performance statistics were carried out. They are Nash and Sutcliffe efficiency (E_{NS}), percentage difference (D), and the regression coefficient (R^2). A calibration and validation of E_{NS}, D, and R^2 for hydrology at $E_{NS}>0.5$, $D≤15\%$, and $R^2>0.6$, respectively, for SWAT model was adopted (Santhi et al. 2001; Moriasi et al. 2007). The three statistical model performance used in the calibration and validation procedure for the soil moisture are presented in Table 2.

Soil Moisture Observation

Five-year satellite-based soil moisture observation data from daily soil moisture data of ESA CCI (2011–2015) and observed soil moisture from the study were obtained. Figure 2 shows the trend in the daily observed soil moisture values during the rainy season (April–July 2017). The highest, lowest, and the mean precipitation are 98.35 mm, 1.98 mm, and 48.64 mm, respectively (Oladapo et al. 2018). An increasing precipitation trend was observed from April to July 2017 in the study area. The increasing soil moisture is likely related to the increasing precipitation between April and July which is the rainy season in the study area. This is largely due to reducing temperature during this period. A short variation in daily precipitation was observed in the region as shown in Fig. 3. Model simulated daily soil moisture values from the sub-basin were compared with the daily observed soil

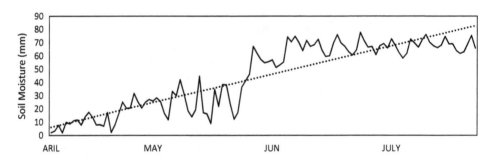

Fig. 2 Trend in the daily soil moisture values observed in 2017 (Oladapo et al. 2018)

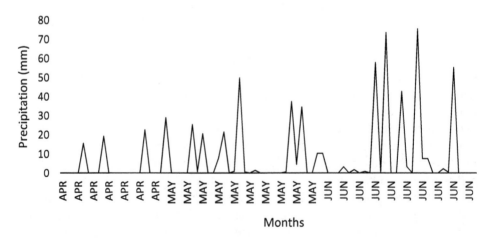

Fig. 3 Precipitation distribution at the watershed during the rainy season in 2017 (Oladapo et al. 2018)

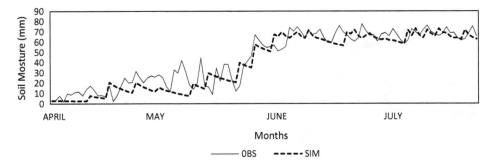

Fig. 4 Comparison between daily means of observations and SWAT simulated soil moisture (April to July 2017) (Oladapo et al. 2018)

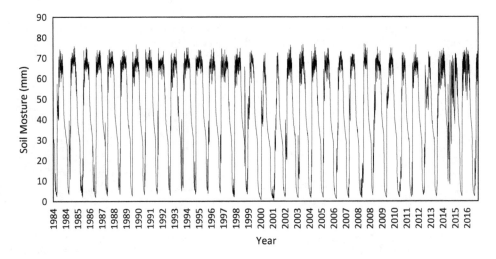

Fig. 5 The 34-year soil moisture simulated by SWAT over Ogbomoso watershed (Oladapo et al. 2018)

moisture values from the field where the sub-basin is located for this four months period. Figure 4 shows that the model closely follows a similar curve to observed value.

Soil Moisture Simulation

The model simulation allows for a 5-year warm-up period. Therefore, a 5-year secondary climate data inputs obtained from NIMET for Ogbomoso metropolis were used for this purpose. The SWAT was used to simulate the soil moisture in the Ogbomoso watershed for 34-year period (1984–2017). Figure 5 shows the temporal variations of soil moisture in the sub-basin of Ogbomoso watershed over the last 34 years. For the 34-year period, the simulated soil moisture declines slightly.

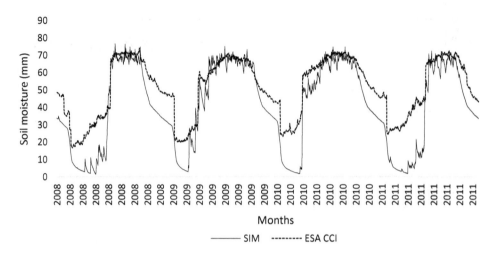

Fig. 6 Comparisons between daily means of SWAT simulated and ESA CCI soil moistures from January to December between 2008 and 2011 (Oladapo et al. 2018)

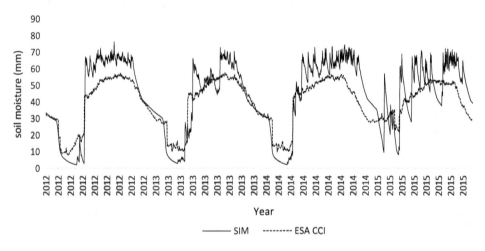

Fig. 7 Comparisons between daily means of SWAT simulated and ESA CCI soil moistures from January to December between 2012 and 2015 (Oladapo et al. 2018)

Figure 6 shows the calibration of the simulated soil moisture with the daily satellite-based soil moisture. There was a good correlation between them. The comparison showed a similar pattern when four years (2012–2015) daily simulated soil moisture from the sub-basin was used for validation as shown in Fig. 7.

The noticeable variation between the daily simulated soil moisture and the satellite-based values of ESA CCI in Figs. 6 and 7 could be traced to the physical and hydraulic parameters associated with the soils, such as soil texture and rainfall, and land cover and land management practices in place. Furthermore, the seasonal variations of the SWAT simulated soil moisture for years 1984, 2000, and 2016 were compared as shown in Fig. 8. Seasonal variations investigated for three years at random showed a similar pattern for the years 1984, 2000, and 2016. The soil

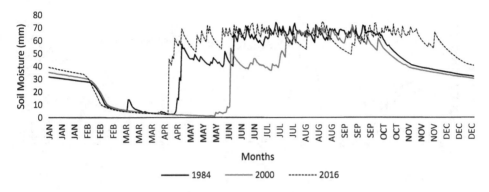

Fig. 8 Seasonal variation of the simulated soil moisture from January to December for year 1984, 2000 and 2016

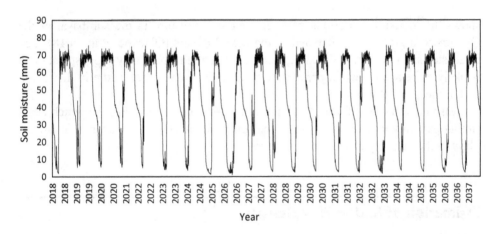

Fig. 9 The 20-year soil moisture forecasted by SWAT over the Ogbomoso watershed (Oladapo et al. 2018)

moisture declines greatly from January till March when the onset of rainfall begins again. This is largely due to reduction in the air temperature of the region. The soil moisture begins to pick up in April being the onset of the raining season and continues to increase till mid-October when it begins to reduce again. The variation in the simulated soil moisture in Fig. 8 is mainly due to the distribution of precipitation in the area.

Forecasted Soil Moisture

The RCP 8.5 projection was used to forecast soil moisture in the region with the assumption of slow growth income and increased population density (Riahi et al. 2011). The RCP 8.5 projection scenario means that there will be increase in the greenhouse gas emission due to the fact that energy is in high demand in the long-term absence of climate change regulations and policies (Riahi et al. 2011). Fig. 9

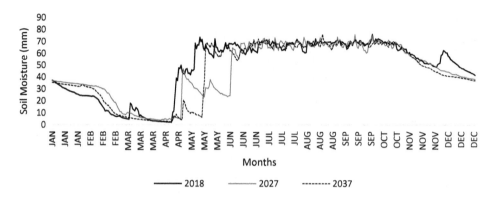

Fig. 10 Seasonal variation of the forecasted soil moisture from January to December for 2018, 2027, and 2037 (Oladapo et al. 2018)

shows the temporal variations of forecasted soil moisture in the sub-basin of the Ogbomoso agricultural watershed in the next 20 years (Oladapo et al. 2018). The forecasted soil moisture showed a general drop in trend which may suggest that there is likely to be a decrease in precipitation and a rise in air temperature in the region within the next 20 years.

Seasonal variation of the SWAT forecasted soil moisture in Fig. 10 which bears similar pattern with that of the simulated soil moisture confirms the key role rainfall plays in the quantity of soil water reserve.

Estimation of Radon Flux Density

The release of radon gas from the soil can take place either by emanation or through the process of transport. The process whereby a ^{222}Rn atom escapes from ^{226}Ra-bearing grains to the pore space in soil is called emanation. The process of radon transport, which occurs in the pore space of the soil, involves the process of advection and diffusion. Diffusion is a product of slope in the radon concentration. Meanwhile, the pressure difference between the ground surface and pore space gives rise to advection. In this chapter, the process of diffusion alone will be considered. This is because influence of most meteorological parameters on the advection can be assumed to be negligible when averaged over one month. Provided that soil under investigation is homogeneous, a one-dimensional diffusion equation can be used to derive the radon flux density from the soil within the region (Goto et al. 2008). There is usually a rise in the radon released into the atmosphere when the soil moisture is low leading to an increase in radon flux density from the soil surface. However, when a certain soil moisture is reached, the radon diffusion and advection process is reduced. This is because the pore space being filled with water slows down the radon flux density from the soil surface (Koarashi et al. 2000; Iimoto 2002). The result of this work is similar to other works on the relationship between soil moisture and radon flux density from the earth surface (Koarashi et al. 2000; Iimoto 2002). The radon flux density values estimated in this work will serve as reference since radon flux data are scarce around the world.

The influence of the soil moisture and soil temperature on the radon flux density is considered. The radon flux density, from the soil surface, was estimated using the UNSCEAR radon flux formula (Zhou et al. 2005).

$$F = R\rho_b \, \varepsilon \left(\frac{T}{273}\right)^{0.75} \sqrt{\lambda Dop} \exp\left(-6Sp - 6S^{14p}\right) \qquad (1)$$

where R is the soil ^{226}Ra content (Bq kg_1), ρ_b is the soil bulk density (kg. m^3), ε is the emanation coefficient of ^{222}Rn in soil, which is a function of the soil temperature (T, in Kelvin scale) and the water saturation fraction (S), λ is the ^{222}Rn decay constant (s^{-1}), p is the soil porosity, and D_o is the ^{222}Rn diffusion coefficient in air (1.1 × 10^{-5} m^2·s^{-1}). The emanation coefficient (ε) of ^{222}Rn in soil was estimated from the fitted formula of Zhuo et al. (2005).

The values of the soil moisture and soil temperature simulated by SWAT were inputted into the equation. The evaluation result showed that the UNSCEAR radon flux formula performed well in estimating the radon flux density in the study area when calibrated with observed atmospheric radon concentration in the study area. The values of the estimated radon flux density ranged between 4.03 and 37.30 mBqm^{-2}·s^{-1} with a mean of 15.09 mBqm^{-2}·s^{-1}. The overall average radon flux density in the area lies below the world mean value (UNSCEAR 2000). The mean value of radon flux density in the area indicated that Ogbomoso metropolis is not prone to risk of radon exposure and may not cause any health risk such as lung cancer and other respiratory tract diseases. For this reason, it is safe to conclude that the radon flux density will not have any radiological health risk to the inhabitants of Ogbomoso metropolis. However, field measurements of the radon flux density within and outside of Ogbomoso are needed to further verify the reliability of these estimates. Since there exist insufficient radon flux measurements around the globe, these set of estimated radon flux densities in Ogbomos will serve as baseline data for future references. Note that the estimated radon flux densities within Ogbomoso investigated in this chapter are between the soil and the atmosphere, hence the health risk is very negligible. The health effect of radon exhalation within buildings in Ogbomoso which is expected to be higher may, however, pose some radiological health risk. For this reason, it is important to review the health implication of radon exhalation values which are triggered by climate change within Ogbomoso.

Conclusion

The effects that climate change has on the variation of soil moisture are the focus of this chapter for the purpose of estimating the health effect of radon flux density within Ogbomoso. A 34-year simulation of soil moisture content was executed within Ogbomoso watershed using the hydrological model, Soil Water Assessment

Tool (SWAT). A 20-year forecast of the soil moisture was also done. The calibrated and validated SWAT model performed well for the simulation of daily soil moisture. A general decline in trend was observed for both the simulated and forecasted soil moisture. This decline in trend may be largely due to rainfall reduction and temperature rising in the region. The SWAT model has proven to be suitable for simulating and forecasting soil moisture in the region. The radon flux density for the study area was estimated using the UNSCEAR radon flux formula. The mean value of the radon flux density of 15.09 mBqm^{-2}·s^{-1} falls below the estimated world average of 33 mBqm^{-2}·s^{-1} by UNSCEAR. The result of the estimated radon flux density showed that Ogbomoso region is not prone to high risk of radon exposure to the public. In other words, the estimated radon flux density in the study area will not pose any radiological health risk such as lung cancer or any other respiratory tract diseases to the inhabitant of Ogbomoso metropolis.

Acknowledgments This research is supported by funding from the UK's Department for International Development (DfID) under the Climate Impacts Research Capacity and Leadership Enhancement (CIRCLE) program implemented by the African Academy of Sciences and the Association of Commonwealth University.

References

Aburnurad KM, Al Tamimi M (2001) Emanation power of radon and its concentration in the soil and rocks. Radiat Meas 34:423

Adabanija MA, Afolabi OA, Olatubosun AT, Kolawole LL (2014) Integrated approach to investigation of occurrence and quality of groundwater in Ogbomoso North, Nigeria. Environ Health Sci 73(1):139–162

AghaKouchak AA (2014) Baseline probabilistic drought forecasting framework using standardized soil moisture index: application to the 2012 United States drought. Hydrol Earth Syst Sci 18:2485–2492

Ajibade AC, Rahaman MA, Ogezi AEO (1988) The Precambrian of Nigeria, a geochronological survey. Publication of the Geological Survey of Nigeria, pp 313–324

Alexander L (2011) Climate science: extreme heat rooted in dry soils. Nat Geosci 4:12

Arnold JG, Srinivasan R, Muttiah RS, Williams JR (1998) Large area hydrologic modeling and assessment, Part I: model development. J Am Water Resour Assoc 34(1):73–89

Ball TK, Nicholson RA, Peachey D (1983) Effects of meteorological variables on certain soil gases used to detect buried ore deposits. Trans Inst Min Metall 92:B183–B190

Brookins DG (1990) The indoor radon problem. Columbia University Press, New York, p 229

Broxton, Xubin Zeng, Damien Sulla-Menashe, Peter A. Troch, (2014) A Global Land Cover Climatology Using MODIS Data. J Appl Meteorol Climatol 53 (6):1593–1605

Chen T, De Jeu R, Liu Y, Van der Werf G, Dolman A (2014) Using satellite based soil moisture to quantify the water driven variability in NDVI: a case study over mainland Australia. Remote Sens Environ 140:330–338

Feng H, Liu Y (2015) Combined effects of precipitation and air temperature on soil moisture in different land covers in a humid basin. J Hydrol 531:1129–1140

Fields RW (2010) Climate change and indoor air quality. Contractor report prepared for U.S. Environmental Protection Agency, Office of Radiation and Indoor Air. pp 1–15

Gassman PW, Reyes MR, Green CH, Arnold JG (2007) The Soil and Water Assessment Tool: historical development, applications, and future research directions. Trans ASABE 50(4):1211–1250

Goto M, Moriizumi J, Yamazawa H, Iida T, Zhou WH (2008) Estimation of global radon exhalation rate distribution. In: The Natural Radiation Environment. 8th International symposium (NRE VIII). America Institute of Physics, New York, pp 169–171

Gupta ML, Douglass AR, Kawa R, R., Pawson. S. (2004) Use of radon for evaluation of atmospheric transport models: sensitivity to emissions, Tellus Series B. Chem Phys Meteorol 56:404–412

Iida T, Ikebe Y, Suzuki, k., Ueno, k., Wang, Z., Jin, Y. (1996) Continuous measurements of outdoor radon concentrations at various location in East Asia. Environ Int 22(suppl. 1):S139–S147

Iimoto T (2002) Environmental factors varying Rn-220 exhalation rate. In: Proceedings of the Third Workshop in Environmental Radioactivity. Tsukuba, pp 189–194, 5–7 Mar 2002. [in Japanese]

International Commission on Radiological Protection (1993) Protection against Radon-222 at home and at work, vol 23. ICRP Publication 65, Annals of the ICRP. Pergamon Press, Oxford

Koarashi J, Amano H, Iida T (2000) Development of model for water and 222Rn in unsaturated soil. In: Proceedings of the 10th International Congress International Radiation Protection Association, Hiroshima, May 14–19

Koster RD, Dirmeyer PA, Guo Z, Bonan G, Chan E, Cox P, Gordon C, Kanae S, Kowalczyk E, Lawrence D (2004) Regions of strong coupling between soil moisture and precipitation. Science 305:1138–1140

Kritz MA, Ronsner SW, Stockwell DZ (1998) Validation of an off-line three dimensional chemical transport model using observed radon profiles. Int J Geophys Res 103:8425–8430

MacDonald AM, Davies J (2000) A brief review of groundwater for rural water supply in sub-Saharan Africa. British Geological Survey Technical Report. W.C./00/33. Natural Environmental Research Council, Nottingham, UK. pp 1–24

Mazur D, Janik M et al (1999) Measurements of radon concentration in soil gas by CR-39 detectors. Radiat Meas 31:295

McColl KA, Alemohammad SH, Akbar R, Konings AG, Yueh S, Entekhabi D (2017) The global distribution and dynamics of surface soil moisture. Nat Geosci 10:100

Moriasi DN, Arnold JG, VanLiew MW, Bingner RL, Harem RD, Veith TL (2007) Model evaluation guidelines for systematic quantification of accuracy in watershed simulations, T. ASABE 50:850–900

Neitsch SL, Arnold JG, Kiniry JR, Williams JR (2005) Soil and water assessment tool (SWAT), theoretical documentation. Blackland Research Center, Grassland, Soil and Water Research Laboratory, Agricultural Research Service, Temple

Ohkura T, Yamazawa H, Moriizumi J, Hirao S, Guo Q, Tohjima Y, Iida T (2009) Monitoring network of atmospheric radon-222 concentration and backward trajectory analysis of radon-222 concentration. J Jan Soc Atmos Environ 44(1):42–51. (in Japanese with English abstract)

Oladapo OO, Amekudzi LK, Oni OO, Aremu AA, Osei MA (2018) Simulation and forecasting of soil moisture content variability over Ogbomoso agricultural watershed using the SWAT model. Official conference proceedings of the European conference on sustainability, energy and environment. *The Internation Academy Forum.* pp 159–173

Ouyang W, Wu Y, Hao Z, Zhang Q, Bu Q, Gao X (2018) Combined impacts of land use and soil property changes on soil erosion in amollisol area under long-term agricultural development. Sci Total Environ 613:798–809

Riahi K, Rao S, Krey V, Cho C, Chirkov V, Fischer G, Kindermann G, Nakicenovic N, Rafaj P (2011) RCP8.5-A scenario of comparatively high greenhouse gas emissions. Climate Change 109:33–57

Santhi C, Arnold JG, Williams JR, Dugas WA, Srinivasan R, Hauck LM (2001) Validation of the SWAT model on a large river basin with point and nonpoint sources. J Am Water Resour Assoc 37:1169–1188

Schery SD, Whittlestone S, Hart KP (1989) The flux of radon and thoron from Australian soils. J Geophys Res 94:8567–8576

Sheffield J, Wood EF (2008) Global trends and variability in soil moisture and drought characteristics, 1950–2000, from observation-driven simulations of the terrestrial hydrologic cycle. J Clim 21:432–458

Sterling SM, Ducharne A, Polcher J (2013) The impact of global land-cover change on the terrestrial water cycle. Nat Clim Chang 3:385

Tanner AB (1980) Radon migration in the ground: a supplementary review. In: TF Gesell, WM Lowder (eds) Natural radiation environment III, Symposium proceedings, Houston, pp 5–56

United Nation Scientific Committee on the Effects of Atomic Radiation (1993) Sources and effects of ionizing radiation. UNSCEAR 1993 report to general assembly, with scientific annexes, 18. United Nation, New York

United Nation Scientific Committee on the Effects of Atomic Radiation (2000) Sources and effects of ionizing radiation. UNSCEAR 2000 report to general assembly, with scientific annexes, volume I Sources, Annex B. United Nation, New York, pp 115–116

Wanders N, Bierkens MF, deJong SM, deRoo A, Karssenberg D (2014) The benefits of using remotely sensed soil moisture in parameter identification of large-scale hydrological models. Water Resour Res 50:6874–6891

Wang T, Ding AJ, Blake DR, Zahorowski W, Poon N, Li YS (2004) Chemical characterization of the boundary layer outflow of air pollution to Honk Kong during February–April 2001. J Geophys Res 108(D20):8787

Wang Y, Yang J, Chen Y, De Maeyer P, Li Z, Duan W (2018a) Detecting the causal effect of soil moisture on precipitation using convergent cross mapping. Sci Rep 8:12171

Wang Y, Yang J, Chen Y, Wang A, De Maeyer P (2018b) The spatiotemporal response of soil moisture to precipitation and temperature changes in an arid region, China. Remote Sens 10:468

Zahorowski W, Chambers CD, Henderson-Sellers A (2004) Ground based Radon-222 observation and their application to atmospheric studies. J Environ Radioact 76:30–33

Zhang K, Wan H, Zhang M, Wang, b. (2008) Evaluation of the atmospheric transport in a GCM using radon measurements; sensitivity to cumulus convection parameterization. Atmos Chem Phys 8:2311–2832

Zhuo W, Iida T, Furukawa M, Guo Q, Kim YS (2005) Soil radon flux density and outdoor radon concentration in East Asia. Elsevier international congress series, vol 1276, p E-285

7

Global Strategy, Local Action with Biogas Production for Rural Energy Climate Change Impact Reduction

A. S. Momodu, E. F. Aransiola, T. D. Adepoju and I. D. Okunade

Contents

Introduction
Literature Review
Methodology
    Digester Design and Development
    Up-Scaling Biogas Production from Laboratory Experiment
    Avoided Emission Calculations/Climate Change Impact Reduction
    Estimation of Life Cycle Costs and Return Streams
    Business Model Formulation
Results and Discussion
    Scaled Up Biogas Production
    Avoided Emissions
    Investment Cost and Variable Cost Stream
    Life Cycle Costs and Return Streams .
    Business Model Formulatio
Conclusions and Recommendation
Reference

Abstract

Global climate change impact is predicted to affect various sectors including the energy demand and supply sectors respectively. Combating this impact will require adoption of both global strategy and localized actions. The use of low

A. S. Momodu (✉)
Centre for Energy Research and Development, Obafemi Awolowo University, Ile-Ife, Nigeria

E. F. Aransiola · T. D. Adepoju · I. D. Okunade
Department of Chemical Engineering, Obafemi Awolowo University, Ile-Ife, Nigeria

carbon strategy based on renewables is a global strategy, while waste management of biodegradable materials through the use anaerobic technology to meet energy demand is a local action. Nigeria is among the vulnerable countries to global climate change impact; this is even more aggravated by its dependence on fossil fuel usage as well as poor waste management, which two, contribute significantly to greenhouse gas emissions. This chapter presents analysis of purified compressed biogas production, a waste conversion option, as a local action to meet rural household energy demand and contribute to global strategy of reducing climate change impact. It discusses both technical and business model approaches to upscale a laboratory experimental procedure for biogas production through anaerobic digestion using vegetal wastes. It shows that using anaerobic technology can achieve efficient waste management and at the same time generate energy that can be used to achieve avoided emissions for climate change impact reduction. The study also concludes that upscaling the project will be sustainable for rural energy augmentation as it produces clean and renewable energy, reduces the use of fossil fuels, provides jobs for skilled and unskilled labor, and generates new return streams.

Keywords

Global strategy · Local action · Waste management · Anaerobic technology · Rural households · Avoided emissions · Business model

Introduction

Energy is a very crucial input for attaining sustainable development. This has become even more pronounced for growth-driven economies, as those in developing countries as Nigeria. Most energy inputs driving economic activities are currently derived from fossil origin, principally hydrocarbon and solid fuels as firewood. These are not socially, environmentally, and economically efficient and therefore not sustainable. Fossil fuels contribute significantly to greenhouse emissions (Bölük and Mert 2014), while solid fuels produce indoor air pollution (WHO 2015). On the other hand, waste is produced from unwanted and discarded materials of human society (Bharadwaj et al. 2015), which needs to be managed properly. Management of waste is a global phenomenon (Isiaka 2017), although the challenge is more pronounced in developing countries for three reasons. These are the increased generation of waste due to increased population, improved living standards, technical and human capacity limitations (Guerrero et al. 2013). African countries are now faced with huge amounts of municipal waste, which has a direct effect on human health, safety, and environment (Bello et al. 2016). Although there is a paucity of data on waste generated annually in Nigeria, however, Isiaka (2017) reports that about 3 two million tons of waste are generated annually, while Vanguard (2017) reports a value of 24 million tons. Of these, only about 20–30% is collected with vegetal wastes, making up 65% of those collected (Isiaka 2017; Ogwueleka 2009).

Vegetal matter refers to substances produced by plant or growing in the manner of plant that can be decomposed by microorganisms. Significant contribution to greenhouse gases and volatile organic compounds emissions comes from their high moisture, organic contents, and biodegradability (Sridevi and Ramanujam 2012). Nonetheless, converting this resource properly could contribute to sustainable energy provision, which Nigeria is in dire need of. Anaerobic digestion is a technology that is recognized to be useful for converting vegetal and animal wastes to generate renewable energy in the form of biogas and organic fertilizer (Arsova 2010). This could result in saving the environment from further degradation, supplementing the energy needs of the rural population (Ahmadu et al. 2009), and generate extra returns streams for farmers and investors (Twidell and Weir 2005; Aransiola et al. 2014; Budzianowski and Brodacka 2017). This contributes significantly to the circular economy in rural communities (Jun and Xiang 2011). Vegetal matters usually occur in large quantities, making it difficult to dispose of easily; over time, this tend to make them become a source of harmful and offensive substances in landfills due to their decomposing qualities (Misi and Forster 2002).

The deployment of a co-digested anaerobic process using cow rumen as inoculum as local action for climate change impact reduction is reported. Essentially, the study involves the design and fabrication of stainless steel farm tanks for anaerobic bio-digestion; upscale of laboratory experimental data for feedstock loading to produce biogas from the substrate; estimate of avoided emissions compared to other fuel sources for cooking energy; evaluate life cycle costs and return streams of biogas produced based on different scenarios; and formulate a business model.

Literature Review

In biogas production, specificity of the substrate collected usually has effects on the type of digester to be used (Christy et al. 2013). Furthermore, biodigester systems are classified based on the method of feeding and number of reactors, with the method of feeding being either batch or continuous, while the number of reactors could also either be a single stage or multi stage (Brown 2006).

The single-stage process involves three stages of the anaerobic process occurring in one reactor. In this process, fermentative bacteria rate of growth is faster than that of acetogenic and methanogenic bacteria (Brown 2006). The consequence is that it results in acid accumulation, pH falling, and methanogenic bacteria growth. For the multi-stage processes, two or more reactors are used to space out the acetogenesis and methanogenesis stages that serve as filters in the system and thus enhancing the digester's production efficiency (Manyi-Loh et al. 2013).

During batch experimental setup, feeding of the digester is only done once, which is at the beginning of the reaction, with the products of the digester being collected at the end of each cycle. However, in that of the continuous system, the feeding and discharging of organic material is continuous (Levenspiel 1999). Rajendran et al. (2012) reviewed different household digesters which are commonly used, reporting that fixed dome (cylindrical digesters) are most commonly used type of digester in

China, while the floating drum digesters are those highly recognized in India, with other digester types used being majorly tubular such as portable plug flow digesters. In terms of features, fixed dome digesters made up of the feed and digestate pipes, a fermentation chamber and a fixed dome on top of the biogas storage, while floating drum digesters are recognized for the floating drum to be at the top of the digestion chamber which separates the gas production and discharge (Neba et al. 2020). Plug flow digesters, which can operate as a household or an industrial digester, have a simple flow pattern without back-mixing. The presence of serious floating tendencies may cause clogging of the flow and prevent the escape of produced biogas in plug flow digesters, though, an inclined plug flow digester at an angle 45° is a better option (Ziyan and Xiaohua 2014; Rajendran et al. 2012). A major advantage the plug flow digester has over fixed dome and floating drum digesters is that there is no difficulty in moving an installed digester, as it is able to produce biogas at a variable pressure and constant volume. In addition, plug flow digesters are capable of managing waste with the range of 11–13% solid concentration (11–13%) (Roos et al. 2004), being always operated in the mesophilic temperature range (Krich et al. 2005).

Global climate change impact is predicted to affect energy sector as with other key sectors of the economy (Arent et al. 2015). One of the strategic approaches to addressing the future of climate change impact is the use of low-carbon energy sources (Yadoo and Cruickshank 2012). Renewables have been taunted as a major contributor to this strategy (Solaun and Cerdá 2019). Developing countries like Nigeria are quite vulnerable to climate change impact, with energy sector being one of the most vulnerable (Ogundipe et al. 2014). Schaeffer et al. (2012) did an extensive review of the vulnerability of the energy sector to climate change. The paper reported on the vulnerability of biofuels to climate change. In addition, as regards biogas production in rural setting, there are ranges of challenges to be considered, which are well documented in literature (Madriz-Vargas et al. 2018). These challenges can affect the operational outcome of renewable energy (RE) technologies as well as the sustainability of the project as a whole. These issues raise questions as regards the set of community capabilities required, appropriate project design, and enabling an external environment for sustainable community RE (CRE) projects.

To introduce the use of biogas as a low carbon strategy to energy generation in the rural areas, the use of user-centered design concept (Redström 2006) is introduced. This approach was proposed as a means to scale-up the process of developing biodigesters in rural areas for energy generation. The essence of the concept is that it takes user experiences into consideration (Redström 2006) in its design. Thus, the basic concept of the business model formulation of the RE power system is to make it be owned, operated, or maintained by a community organization. With this, technical and nontechnical problems such as the issue of social integration of RE technologies, lack of investment and maintenance capabilities, as well as end-user education (Madriz-Vargas et al. 2018; Margolis and Zuboy 2006) are eliminated.

Methodology

This section describes the approach adopted to design and fabricate the digesters, upscaling the substrate used to feed the digesters from a laboratory experiments (Adepoju 2019) conducted. The scaling up includes technical design and development of the biodigester, up-scaling biogas production from laboratory experiment, evaluation of avoided emissions, estimation of life cycle costs as well as return streams, and business model formulation.

Digester Design and Development

In the design, fabrication, and construction of the plug flow biodigester, the factors that affect the building of digester for optimum biogas yield (Jiang et al. 2011) should be taken into consideration. The principal materials to be selected for the fabrication will be stainless steel sheets because of its durability and as well as its ability to absorb heat easily, which improves mesophilic anaerobic digestion as to when compared to cement and block. The design analysis includes the design specification of the biodigester and the length of the digester is 5.96 m. The required length-to-width will be within the ratio of 3.5:1 based on the Natural Resources Conservation Service (2004) length-to-width ratio for manure in plug-flow digesters. Therefore, the dimensions of the digester are:

Length of the digester (L) = 5.96 m
Width of the digester (W) =0.71604 m
The volume of the digester = 9.6 m^3

A plug flow reactor has the following components as shown in Fig. 1:

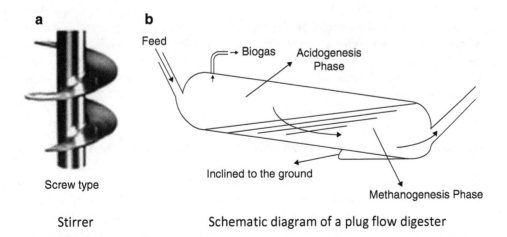

Screw type

Stirrer   Schematic diagram of a plug flow digester

Fig. 1 Schematic diagram of a plug flow digester

Table 1 Theoretical comparison of different substrates to meet energy needs of 100 households

Substrates	Biogas potential (L/kg)[a]	Expected daily biogas production needed for 100 households (m^3)	Expected biomass (waste) for feeding (tonnes)	Expected size of biodigester (m^3)
Vegetal matter	70	39.13	133.91	67
Corn silage	200	39.13	465.83	23
Grass silage	220	39.13	423.48	21

[a]Source: Arora and Linton (2011)

1. Inlet and outlet pipes: The inlet pipe can also be referred to as the feed pipe while the outlet pipe as the digestate pipe. The pipes to be used will be of a steel rod with diameters of 500 cm and 250 cm as the inlet for feeding the substrate into the digester and outlet pipes for discharging the consumed slurry at the end of the digestion, respectively. The inlet pipe which will be at an angle of 45° for convenient channel of the substrate into the digester and the entry will be closed to prevent air from getting to the residue for easy break down of waste materials in the digester.
2. Stirrer: A screw-type, stainless material stirrer is shown in Fig. 1. The stirrer will be connected to a motor to drive it in order to create a turbulent motion of the substrates.
3. Storage tank: The storage tanks will be balloons made from high density polyethylene (HDP) or floating tank made from fiber glass for collection of biogas produced.
4. Effluent collection tank: The effluent (digestate) will be collected in a plastic tank.

The design involves the construction of seven 9.6 m^3 biodigesters. A combination of the seven farm tanks biodigesters was designed to produce approximately 39.13 m^3 of biogas daily. It is also expected that for each biodigester, about 64% of the biodigester size would be for gas accumulation. Table 1 shows different substrates with their biogas potentials, expected daily biogas production needed for 100 households, expected biomass feedstock, and expected size of biodigester. The use of grass silage as substrate for biogas production required the least sized digester at 21 m^3.

Up-Scaling Biogas Production from Laboratory Experiment

This section describes up-scaling the steps to the pilot scale from laboratory experiment. The laboratory experiment involves the use of portable 20 l plastic containers, modified as digesters, as shown in Fig. 2. Gas collection was done through water displacement method (Otun et al. 2015). Stirring was done by shaking the biodigester to prevent thickening and settling of the slurry. The experiment consists of the use of fresh waste samples of vegetables and fruits serving as feedstocks, and

Fig. 2 Laboratory-scale experiment (Adepoju 2019)

cow rumen fluid used as inoculum, collected from a food market and an abattoir. The vegetables and fruits, made up of watermelon, tomatoes, and oranges, were grinded to increase the surface area. The cow rumen fluid collected in polyethylene bags was stored at room temperature for 4 days (Elhasan et al. 2015). These were mixed with water to form slurry in the biodigester (Budiyono et al. 2014). Fresh vegetal wastes (V) mixed with cow rumen fluid (R) and clean water (W) in different V:W:R ratio and batch fed into the digester. The retention time was 30 days.

The biodigester was scaled up based on the data acquired from the laboratory-scale experiment to meet the energy needs of about a hundred (100) rural households. The energy cooking need was estimated based on Riuji (2005) and Rea (2014) that 100 l of biogas will produce 23 min of cooking time. It was assumed that a household of 5 will use 390 l of biogas for 90 min cooking time in a day. With these data, it was estimated that 100 households will require 39,130 l of biogas daily. According to Arora and Linton (2011), 1 kg of vegetal waste will produce on average 2.33 l of biogas daily and a total of 70 l biogas over a retention time of 30 days. From Arora (2011), to achieve a daily production rate of 39,130 l of biogas, will require 16,670 kg of vegetal waste without the use of inoculum, requiring an organic loading rate (OLR) of 555.67 kg daily. With the use of cow rumen as inoculum, the efficiency of the substrate is improved by 26% (Stan et al. 2018), reducing the amount of vegetal matter needed to 13,310 kg or an OLR of 443.67 kg daily. The size of the biodigester is dependent on the amount of waste needed. Using the laboratory experiment that requires 4 kg of waste for 0.02 m^3 of biodigester, the equivalent biodigester size for 13,310 kg of waste was estimated to be 67 m^3 to produce 39,130 l of biogas daily to meet the energy needs of ~100 rural households.

Avoided Emission Calculations/Climate Change Impact Reduction

The avoided emissions were estimated based on the biogas equivalent to the fossil fuel conversion method (B-Sustain 2013b), and 1 m^3 of biogas equivalent for each of the fuels is given in Table 2. The emission factors of biogas, kerosene, LPG, kerosene, and firewood were obtained from Simon et al. (2006), which was used for the estimation of the CO_2 emission. The CO_2 emission reduction potential of using biogas in relation to other fuels was evaluated by subtracting the emission from the particular fuel and that from biogas.

Estimation of Life Cycle Costs and Return Streams

In order to estimate the unit cost for the produced biogas for either cooking or electricity generation will involve life cycle analysis (Lakhani et al. 2014). For this project, a 20-year life cycle was assumed for biogas generation (Tsaganakis and Papadogiannis 2006). Based on the assumption, this life cycle cost was calculated thus:

$$Life\ cycle\ cost\ (LCC) = \frac{Total\ cost}{Energy\ derived}$$

where Total Cost = Fixed Costs + Variable Costs.

1. **Cooking**: To calculate LCC for energy derived from biogas for cooking, the assumption made is that the digester has a life cycle of 20 years and production capability of 39.13 m^3 per day. It is also assumed that the digester will work for 300 days in a year.
2. **Electricity generation**: For the energy derived from biogas for electricity, the assumption made for the life cycle is 20 years, production capacity of 849845.57 BTU/day, and 300 days of yearly operation.
3. **Digestate production**: To calculate LCC for digestate from the biodigester, the assumption made is that the digester has a life cycle of 20 years and production capability of 54.26 kg per day, with 300 working days per year.

Table 2 Biogas equivalent to fossil fuels and firewood

	Fuel	Quantity (kg)
	LPG	0.45
	Kerosene	0.6
1 m^3 of biogas equivalent to	Firewood	3.5
	Furnace oil	0.4
	Petrol	0.7
	Diesel	0.5

Source: B-Sustain (2013b)

The life cycle costs obtained were used to calculate the sales price based on the value added tax (VAT) of 5%, bank interest of 25% (assuming bank loan was secured to execute the project), sales tax of 5% (assuming sales tax will be charged), and a profit margin of 5%, which gives a sum of 40%. The sales prices were estimated by calculating 40% of the life cycle costs.

Business Model Formulation

The user-centered approach represents the concept behind the business model formulation to enable coupling of technical and economic aspects of biogas production principally for rural energy supply. This also includes issues of cash flows, access to finances, and fuel switching in rural areas of Nigeria.

The formulated business model gives potential investors an overview of typical rural energy markets, target customers, and potential return streams to be earned from a biogas production business. The adapted model is an integrated one with three domains. The first domain is the upstream side or technical input, which includes planning, R&D resources, and installation of the production process of the business that includes building of the biodigester, storage system, laying of supply pipelines, and sourcing for raw materials to be used from farms and food industries biowastes. The second domain reflects the transformation of the raw materials from the biodigester to create value for the target market. This involves anaerobic digestion to yield raw biogas and the byproduct as digestate (fertilizer), piping the biogas to connected households for cooking, and subsequently cleaning and pressurizing the biogas for electricity generation to the target customers. The last domain is the market segment of the business. This involves the marketing of biogas and digestate to target customers which include: cottage industries, residential use, commercial, and farmers.

Smallholders in rural areas usually have limited access to finance as they have to confront different challenges from bank demands, such as the complex and drawn-out procedure of documentations, high bank charges, short-term nature of the credit, and disturbing problem of mortgage for security (Abdullah et al. 2015). In the same vein is the issue of fuel switching. The Nigeria household energy mix is as shown in Fig. 3, with firewood making up 56% and sawdust being the least at 2%. To increase the proportion of clean energy sources in the form of gas such as LPG and biogas will require substantial fuel switching. The current proportion of clean energy sources in the household energy mix is barely 5%. This will even be worse in rural areas in Nigeria, where dependence on unclean sources such as firewood and kerosene are even more prevalent. For clean energy sources such as biogas or LPG to be more acceptable, the proposed business model takes into consideration the provision for household fuel switch, particularly in rural areas. The most used fuel stove currently in rural areas of Nigeria is based on firewood that depends on the traditional three-stone stoves or mud-built stoves. These stoves are grossly inefficient and unhealthy for humans and the environment due to the release of inimical particulate matter into the air (Akinbami and Momodu 2013). So tackling fuel

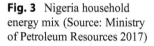

Fig. 3 Nigeria household energy mix (Source: Ministry of Petroleum Resources 2017)

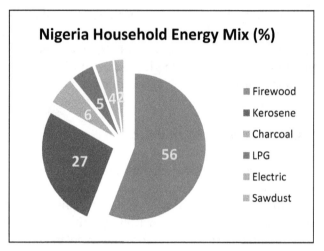

switching for rural households will involve tackling the economics of fuel and stove type, respectively. Other issues to be addressed include but not limited to access conditions to fuels, technical characteristics of cook stoves and cooking practices, cultural preferences, and health impacts (Masera et al. 2000).

Results and Discussion

This section presents the results and analysis of the study. This includes the process of scaling up laboratory experiments to the field scale, avoided emission, investment cost, variable cost stream, lifecycle cost, and return streams, as well as business model formulation.

Scaled Up Biogas Production

Scaling up the biogas production to meet energy provision consists of a system of seven tanks of biodigesters that are each sized at 9.6 m^3. In this farm of biodigesters, the tanks tilted at 45° will consist of a mechanical stirrer, inlet, and outlet for feeding slurry and evacuating digestate, respectively. Biogas collection will be done using gas hoses connected to the gate valve at the top of the biodigester tanks (Arnott 1985). The raw biogas will be subjected to purification, liquefaction, storage, and transportation (Ahmad et al. 2018). The purification could be done either using water scrubber and iron filings or through cryogenic processes, depending on the level of purity required, which is determined by use. The water scrubber and iron filings will be used to clean the biogas by removing CO_2 and H_2S when the gas is needed only for cooking. On the other hand, cryogenic process is used when the gas purity is required at 94–99% level. This is an environmentally friendly biogas upgrading and

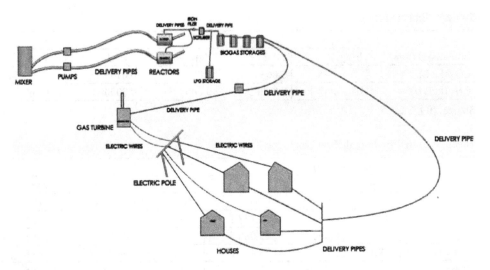

Fig. 4 Schematic layout of biogas production system for household energy and electricity generation

biomethane liquefaction system which separates pollutants and CO_2 from biogas via low-temperature (Ahmad et al. 2018). The cryogenic process will be used to purify and liquefy the biogas to be produced for the reason of storage, transportation, and high energy content gas demand of some applications. Thus, it is imperative that the concentration of methane in the biogas be increased with the removal of CO_2.

The gas will be stored in four interconnected fiber-glass tanks of 10 m³ each. Two fiber glass tanks will supply gas for electricity generation, while the other two fiber glass tanks will be for cooking. For safety measures, there will be an inclusion of a water-cooling system for the storage tanks and regular cleaning of the pipe holes to avoid blockages and leakages and the use of several pressure relief valves along the pipeline to control the pressure of the gas. A schematic of the biogas production system is depicted in Fig. 4.

Avoided Emissions

Utilization of energy derived from biogas technology operates to reduce GHG emissions, particularly CO_2, by reducing the demand for fossil fuels and waste management (B-Sustain 2013a). The emission factors used for estimating the GHG emission from different fuel types are shown in Table 3, while Table 4 shows the total GHG emission of different fuel types. Table 5 shows that using 39.13 m³/day of biogas produced from vegetal matter could reduce the CO_2 emission of 9.69 kg from LPG, 37.49 kg from kerosene, 23.15 kg from diesel, and 181.65 kg from firewood use. Therefore, using biogas produced from waste of vegetal matter instead of LPG, kerosene, diesel, and firewood fuels is a means to mitigate the environmental impacts of CO_2 and other GHG.

Table 3 Emission factors

Emission factors	Fuel type				
	Biogas	LPG	Kerosene	Diesel	Firewood
CO_2 (kg/MJ)	0.055	0.063	0.072	0.074	0.11
CH_4 (kg/MJ)	0.000001	0.000001	0.000003	0.000003	0.00003

Source: (IPCC 2006)

Table 4 Emissions from different fuel types

	Gas emitted (Kg)					
	Daily		Annually		For a period of 20 years	
Fuel type	CO_2	CH_4	CO_2	CH_4	CO_2	CH_4
Biogas	40.6	0.00074	12,178.1	0.22	243,560.8	4.5
LPG	50.3	0.00079	15,086.5	0.24	301,729.3	4.8
Kerosene	78.1	0.0032	23,424.2	0.98	468,484.6	19.6
Diesel	63.7	0.0026	19,122.3	0.77	382,444.9	15.5
Firewood	222.2	0.06	66,673.2	17.86	1,333,466	357.2

Table 5 Avoided emissions from different fuel types

	Gas emitted (Kg)					
	Daily		Annually		For a period of 20 years	
Fuel Type	CO_2	CH_4	CO_2	CH_4	CO_2	CH_4
Biogas	0	0	0	0	0	0
LPG	9.7	5.3E-05	2908.4	0.016	58,168.55	0.3
Kerosene	37.5	0.0025	11,246.2	0.76	224,923.8	15.1
Diesel	23.2	0.0018	6944.2	0.55	138,884.1	11.0
Firewood	181.7	0.059	54,495.3	17.62	1,089,906	352.7

Investment Cost and Variable Cost Stream

The cost stream, as shown in Table 6, itemizes the cost of materials to be used for the construction of 67 m³ biodigester, including its civil works, the pipelines for gas evacuation and distribution, and the storage tank. The materials include stainless steel, copper pipes for connections, different sizes of valves, scrubber and iron filings, compressor, LPG cylinder, gas turbine, storage tanks, single burner cooking stoves, and rubber hoses. For purer gas needs, the materials will include cryogenic equipment.

Life Cycle Costs and Return Streams

Based on the production of biogas from the biodigesters, life cycle costs (LCC) were estimated for cooking, electricity generation, and digestate production in three different scenarios. The first scenario, called A, considered the production of

Table 6 Cost stream

Item description	Estimated unit cost	Estimated quantity	Amount (₦)
Fixed costs			
Stainless steel 9.6 m³ tanks	1000,000	7	7,000,000
Different sizes of connecting tubes	500	160	80,000
Different sizes of valves	5000	160	800,000
Cryogenic equipment	3,500,000	2	7,000,000
Land cost and site construction	1,500,000	4	6,000,000
Transportation	2,000	12	24,000
Storage tanks	200,000	18	3,600,000
Average labor (skilled and unskilled) costs	8,000	20	160,000
Purchase of pressure relief valves	1,000	10	10,000
Purchase of copper pipes	6,000	2	12,000
Construction of biogas burner cooking stoves	8,000	100	800,000
Gas distribution lines	1000	500	500,000
			25,986,000
Variable cost			
Substrate collection	399,286	1	399,286
Sample analysis	8000	1	8,000
Operation and maintenance costs	2,000,000	1	2,000,000
Salvage cost	700,000	1	700,000
Total			**29,093,285.60**

digestate and biogas for cooking only, in which the LCC are $0.13/kg (₦ 44.67/kg) and $0.14/m³ (₦50.50/m³), respectively. The second scenario B is the production of the digestate and biogas for electricity, while the last scenario C considers the production of digestate, 50% biogas for cooking and 50% biogas for electricity. These life cycle costs enable the cost comparison analysis of various scenarios (Table 7).

Theoretically, the investment cost is estimated as ₦29.1 million, while the return streams at different scenarios are the same, which is ₦ 40.74 million with a profit margin of ₦ 11.74 million and an average annual profit of ₦0.59 million. With the positive difference between the LCC and the return stream generated, the biogas project shows good performance with hope the business will allow for its smooth operation to enable quick cost recovery.

Business Model Formulation

The concept proposed for the rural energy biogas business, based on user-centered design, is diagrammatically shown in both Figs. 5 and 6, respectively. Figure 5 shows a simplified schematic diagram of the business model that mimics what is

Table 7 LCC and return stream

Scenario A

	LCC (₦/kg)	Sales price (₦/kg)	Quantity (tonnes)	Revenue (₦) (million)
Cooking	50.50	70.69	288.1	20.37
Digestate	44.67	62.54	325.6	20.37
Total				**40.74**

Scenario B

	LCC	Sales price	Quantity	Revenue (₦) (million)
Electricity generation	0.00285 (₦/BTU)	0.004 (₦/BTU)	5 million BTU	20.37
Digestate	44.67 (₦/kg)	62.54 (₦/kg)	325.6 tonnes	20.37
Total				**40.74**

Scenario C

	LCC	Sales price (₦)	Quantity	Revenue (₦) (million)
Cooking	₦ 67.33/kg	94.25	144.04 tonnes	13.58
Electricity generation	₦ 0.0038/BTU	0.0053	2 million BTU	13.58
Digestate	₦ 29.79/kg	41.70	325.6 tonnes	13.58
Total				**40.74**

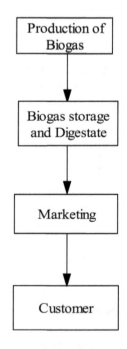

Fig. 5 Simplified schematic diagram of the biogas business model

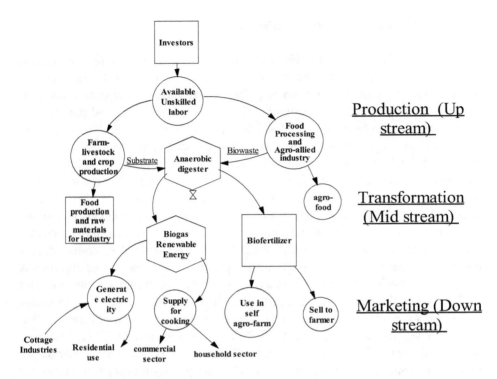

Fig. 6 Schematic diagram of the adopted business model. Source: Adapted from Yousuf et al. (2017)

obtained in traditional vertically integrated electric power systems (Walsh and Todeva 2005). Both diagrams show three distinct operations of technical inputs, storage of biogas and digestate, as well as marketing. The first domain contains most of the investment cost segment of the model. The cost stream in the business model involves investment in livestock and crop farming, food processing, agro-allied industry, and biodigester construction, while the biogas yield and bio-fertilizer will generate the return stream for the business. From the perspective of the investors, the financial feasibility of the biogas business would be assessed based on its return and cost streams as well as its ability to make profit. In Nigeria, rural areas are mostly faced with challenges in fuel switching, access to finances, and cash flows. For this project, in order to make the business model operational, the issue of fuel switching is addressed. Addressing this challenge involves adopting a user-centered approach that combines different types of business models. At inception, a niche business model would be introduced to incentivize the first few households that adopt (early adopters) the use of biogas as an energy source, where a single gas stove burner would be given to them free but have to buy the gas to be used. For other households (mid to late adopters), the razor and blade business model would be introduced, where the users will have to buy the gas stoves at graduated subsidized prices and still also have to purchase the gas for use.

Conclusions and Recommendations

Global climate change impact needs to be addressed strategically using local actions. One of such strategy is the use of low-carbon energy resource to meet energy demand. Energy supply in Nigeria is at best erratic, and this is even more pronounced in the rural areas. In addition, Nigeria, particularly her rural populace, is among the most vulnerable to global climate change impact; this is even more aggravated by Nigeria's dependence on fossil fuel usage as well as poor waste management, which two, contribute significantly to greenhouse gas emissions and influences climate change impact significantly. Using waste conversion option through biogas production to meet rural household energy demand is a local action to contribute to the global strategy. Presented are both technical and business model for using biogas production as a means of waste management measure and energy supply source. The study shows that, theoretically, scaling up of laboratory experiments for biogas production through the process of bio-waste anaerobic digestion is not only possible but also comes with a positive difference in climate change impact reduction as well as between costs and returns. First, the project will achieve avoided emissions to reduce climate change impact, which is put at 58,168.55 kg, 224,923.8 kg, 138,884.1 kg, and 1,089,906 kg, respectively, for LPG, kerosene, diesel, and firewood. In terms of costs and returns, three different scenarios are looked into, namely, Scenario A involves just cooking using the biogas produced for cooking and selling of the digestate as bio-fertilizer; Scenario B, involving using the biogas produced for electricity generation, and selling of the digestate; and Scenario C, cooking, electricity generation, and digestate sales are considered. The total cost estimated for starting the project is approximately ₦29.1 million, and the returns stream is estimated at ₦40.74 million. This gives a profit margin of ₦11.64 million for the 20-year life cycle and an average yearly profit of ₦0.59 million. It is important to note that each of the scenarios presented the same total cost though the LCC and sales price pathways were different. Biogas energy production could be effective to transforming the rural economy. Another aspect of interest that needs to be further investigated is that the CO_2 got from cryogenic processes could be channeled. This will generate some income while also removing CO_2 from the atmosphere. It is further recommended that an actual pilot scale be done with data to verify the theoretical estimation made in this study.

References

Abdullah DZ, Khan SA, Jebran K, Ali A (2015) Agricultural credit in Pakistan: past trends and future prospects. J Appl Environ Biol Sci 5:178–188

Adepoju TD (2019) Biogas generation using vegetal matter via anaerobic digester. A Bachelor's degree thesis. Obafemi Awolowo University, Ile-Ife

Ahmad NE, Mel M, Sinaga N (2018) Design of liquefaction process of biogas using Aspen HYSYS simulation. J Adv Res Biofuel Bioenergy 2:10–15

Ahmadu TO, Folayan CO, Yawas DS (2009) Comparative performance of cow dung and chicken droppings for biogas production. Niger J Eng 16(1):154–164

Akinbami CA, Momodu AS (2013) Health and environmental implications of rural female entrepreneurship practices in Osun state Nigeria. Ambio 42(5):644–657

Aransiola EF, Ojumu TV, Oyekola OO, Madzimbamuto TF, Ikhu-Omoregbe DIO (2014) A review of current technology for biodiesel production: state of the art. Biomass Bioenergy 61:276–297

Arent DJ, Tol RS, Faust E, Hella JP, Kumar S, Strzepek KM, Tóth FL, Yan D, Abdulla A, Kheshgi H, Xu H (2015) Key economic sectors and services. In: Climate change 2014 impacts, adaptation and vulnerability: part a: global and sectoral aspects. Cambridge University Press, Cambridge, pp 659–708. 2010

Arnott M (1985) Biogas/biofertilizer business handbook (no. PB-87-158937/XAB; PC/ICE/R-48). Peace Corps, Washington, DC. Information Collection and Exchange

Arora SM, Linton JA (2011) Mississippi renewable energy and energy efficiency report. A snap shot of related activities in the state of Mississippi (no. DOE-MTA-85002). Mississippi Technology Alliance, Jackson

Arsova L (2010) Anaerobic digestion of food waste: current status, problems and an alternative product. Department of Earth and Environmental Engineering Foundation of Engineering and Applied Science Columbia University

Bello IA, bin Ismail MN, Kabbashi NA (2016) Solid waste management in Africa: a review. Int J Waste Resour 6(2):1–4

Bharadwaj A, Yadav D, Varshney S (2015) Non-biodegradable waste- its impact and safe disposal L. Int J Adv Technol Eng Sci III:184–191

Bölük G, Mert M (2014) Fossil & renewable energy consumption, GHGs (greenhouse gases) and economic growth: evidence from a panel of EU (European Union) countries. Energy 74:439–446

Brown VJ (2006) Biogas a bright idea for Africa. Environ Health Perspect 114:A300–A303

B-Sustain (2013a) Environmental and social benefits of biogas technology. Retrieved from: http://www.bsustain.in/faqs.html. Accessed July 2019

B-Sustain (2013b) Biogas equivalent to fossil fuels and its emission comparison. Retrieved from: http://www.bsustain.in/faqs.html. Accessed July 2019

Budiyono B, Widiasa IN, Johari S, Sunarso S (2014) Increasing biogas production rate from cattle manure using rumen fluid as inoculums. Int J Sci Eng 6(1):31–38

Budzianowski WM, Brodacka M (2017) Biomethane storage: evaluation of technologies, end uses, business models, and sustainability. Energy Convers Manag 141:254–273

Christy EM, Sampson NM, Edson LM, Anthony IO, Michael S (2013) Microbial anaerobic digestion (bio-digesters) as an approach to the decontamination of animal wastes in pollution control and the generation of renewable energy. Int J Environ Res Public Health 10:4390–4417

Elhasan MMAA, Abdalla IF, Ibrahim AAA (2015) Economics of onion production under cooperative and private schemes in Khartoum North, Sudan (No. 15/05). NAF International Working Paper Series

Guerrero LA, Maas G, Hogland W (2013) Solid waste management challenges for cities in developing countries. Waste Manag 33(1):220–232

Isiaka T (2017) Challenges of waste management in Lagos. Susty Vibes

Jiang X, Sommer SG, Christensen KV (2011) A review of the biogas industry in China. Energy Policy 39(10):6073–6081

Jun H, Xiang H (2011) Development of circular economy is a fundamental way to achieve agriculture sustainable development in China. Energy Procedia 5:1530–1534

Krich K, Augenstein D, Batmale J, Benemann J, Rutledge B, Salour D (2005) Biomethane from dairy waste: a sourcebook for the production and use of renewable natural gas in California. Western United Dairymen. Available at: http://www.suscon.org/cowpower/biomethaneSourcebook/Full_Report.pdf

Lakhani R, Doluweira G, Bergerson J (2014) Internalizing land use impacts for life cycle cost analysis of energy systems: a case of californias photovoltaic implementation. Appl Energy 116:253–259

Levenspiel O (1999) Chemical reaction engineering. Ind Eng Chem Res 38(11):4140–4143

Madriz-Vargas R, Bruce A, Watt M (2018) The future of community renewable energy for electricity access in rural Central America. Energy Res Soc Sci 35:118–131

Manyi-Loh CE, Mamphweli SN, Meyer EL, Okoh AI, Makaka G, Simon M (2013) Microbial anaerobic digestion (bio-digesters) as an approach to the decontamination of animal wastes in pollution control and the generation of renewable energy. Int J Environ Res Public Health 10(9):4390–4417

Margolis R, Zuboy J (2006) Nontechnical barriers to solar energy use: review of recent literature (No. NREL/TP-520-40116). National Renewable Energy Laboratory (NREL), Golden

Masera OR, Saatkamp BD, Kammen DM (2000) From linear fuel switching to multiple cooking strategies: a critique and alternative to the energy ladder model. World Dev 28(12):2083–2103

Ministry of Petroleum Resources (2017) National gas policy approved by FEC. Assessed from http://www.petroleumindustrybill.com/wpcontent/uploads/2017/06/National-Gas-Policy-Approved-By-FEC-in-June-2017.pdf. 27 May 2019

Misi SN, Forster CF (2002) Semi-continuous anaerobic co-digestion of agro-wastes. Environ Technol 23(4):445–451

Natural Resources Conservation Service (2004) Anaerobic digester-controlled temperature. 2004Code366 Natural resources conservation service, Conservation practice standard. Available at http://efotg.sc.egov.usda.gov//references/public/CA/366std-9-04.pdf

Neba FA, Asiedu NY, Addo A, Morken J, Østerhus SW, Seidu R (2020) Biodigester rapid analysis and design system (B-RADeS): a candidate attainable region-based simulator for the synthesis of biogas reactor structures. Comput Chem Eng 132:106607

Ogundipe A, Akinyemi O, Alege PO (2014) Energy supply and climate change in Nigeria. J Environ Earth Sci 4:14

Ogwueleka TC (2009) Municipal solid waste characteristics and Management in Nigeria. Iran J Environ Health Sci Eng 6(3):173–180

Otun T, Ojo O, Ajibade F (2015) Evaluation of biogas production from the digestion and co-digestion of animal waste, food waste and fruit waste. Int J Energy Environ Res 3(3):12–24

Rajendran K, Aslanzadeh S, Taherzadeh JM (2012) Household biogas digesters- a review. Energies 5(8):2911–2942

Rea J (2014) Kinetic modeling and experimentation of anaerobic digestion. Doctoral dissertation, Massachusetts Institute of Technology

Redström J (2006) Towards user design? On the shift from object to user as the subject of design. Des Stud 27(2):123–139

Riuji LC (2005) Research on anaerobic digestion of organic solid waste at household level in Dar Es Salaam, Tanzania. Doctoral dissertation, ARDHI University

Roos KF, Martin JB, Moser MA (2004) AgStar handbook: a manual for developing biogas systems at commercial farms in the United States. EPA-430-B-97-015. USEPA, Washington, DC. Available at: http://www.epa.gov/agstar/documents/AgSTAR-handbook.pdf

Schaeffer R, Szklo AS, de Lucena AFP, Borba BSMC, Nogueira LPP, Fleming FP, Troccoli A, Harrison M, Boulahya MS (2012) Energy sector vulnerability to climate change: a review. Energy 38(1):1–12

Simon Eggelston, Leandro Buendia, Kyoko Miwa, Todd Ngara, Kiyoto Tanabe (2006) Guidelines for national greenhouse gas inventories. The Institute for Global Environmental Strategies (IGES) for the IPCC, Tokyo, Japan

Solaun K, Cerdá E (2019) Climate change impacts on renewable energy generation. A review of quantitative projections. Renew Sust Energ Rev 116:109415

Sridevi D, Ramanujam RA (2012) Biogas generation in a vegetable waste anaerobic digester: an analytical approach. Res J Recent Sci 1(3):41–47

Stan C, Collaguazo G, Streche C, Apostol T, Cocarta D (2018) Pilot-scale anaerobic co-digestion of the OFMSW: improving biogas production and startup. Sustainability 10(6):1939

Tsaganakis KP, Papadogiannis C (2006) Technical and economic evaluation of the biogas utilization for energy production at Iraklio municipality, Greece. Energy Convers Manag 47:844–857

Twidell J, Weir T (2005) Biomass and biofuels. In: Renewable energy resources, 2nd edn. Spon, London, pp 351–399

Vanguard (2017) Nigeria generated 24m tonnes of waste more than rice in 18 states in 2016 – reports. Assessed from https://www.vanguardngr.com/2017/08/nigeria-generated-24m-tonnes-waste-rice-18-states-2016-reports/. 16th Aug 2019

Walsh PR, Todeva E (2005) Vertical and Horizontal Integration in the Utilities Sector: the case of RWE. Unpublished paper

WHO (2015) Residential heating with wood and coal: health impacts and policy options in Europe and North America. World Health Organisation, Copenhagen, Denmark

Yadoo A, Cruickshank H (2012) The role for low carbon electrification technologies in poverty reduction and climate change strategies: a focus on renewable energy mini-grids with case studies in Nepal, Peru and Kenya. Energy Policy 42:591–602

Yousuf A, Sultana S, Monir M, Karim A, Rahmaddulla S (2017) Social business models for empowering the biogas technology. Energy sources. Part B Economics, planning and policy. 12:99–109

Ziyan T, Xiaohua L (2014) Design and optimization principles of biogas reactors in large scale applications. In: Reactor and process design in sustainable energy technology. Elsevier, Amsterdam, Netherlands, pp 99–134

8

Climate Change Implications and Mitigation in a Hyperarid Country: A Case of Namibia

Hupenyu A. Mupambwa, Martha K. Hausiku, Andreas S. Namwoonde, Gadaffi M. Liswaniso, Mayday Haulofu and Samuel K. Mafwila

Contents

Introduction ....
Namibia: A Vast Place
Aerosol Monitoring in Namibia and Its Importance
Climate Change and the Oceans: Implications for Namibia
Alternative Water Sources in a Hyperarid Namibia .....
    Availability of Water in Namibia
    Seawater Desalination in Namibia
Alternative Food Sources under a Changing Climate: Mushrooms
    Climate Change Implications on Cultivation of Macrofungi in Namibia
    Current and Future Research on Mushrooms in Namibia

H. A. Mupambwa (✉)
Desert and Coastal Agriculture Research, Sam Nujoma Marine and Coastal Resources Research Centre (SANUMARC), Sam Nujoma Campus, University of Namibia, Henties Bay, Namibia
e-mail: hmupambwa@unam.na

M. K. Hausiku
Mushroom Research, SANUMARC, Sam Nujoma Campus, University of Namibia, Henties Bay, Namibia
e-mail: mkhausiku@unam.na

A. S. Namwoonde
Renewable Energy Research, SANUMARC, Sam Nujoma Campus, University of Namibia, Henties Bay, Namibia
e-mail: anamwoonde@unam.na

G. M. Liswaniso
Mariculture Research, SANUMARC, Sam Nujoma Campus, University of Namibia, Henties Bay, Namibia
e-mail: gliswaniso@unam.na

Organic Agriculture and Its Potential Under a Changing Climate
Conclusions ....
References

Abstract

Namibia is the most arid country in sub-Saharan Africa characterized by the existence of two deserts, the Namib and the Kalahari. However, though being arid, agriculture still plays a critical role in Namibia's economy, which includes both crop and animal production. Furthermore, the country is endowed with vast marine resources, with its marine waters being equivalent to two-thirds of Namibia's terrestrial environment. In the face of climate change and a growing population, there is a need for Namibia to continue with its climate smart efforts which is critical in shifting the country from its current dependency on imports thus increasing the country's food self-sufficiency. This chapter highlights the threats posed by climate change, both on land and the marine environment of the country, which has potential negative impacts on the economy. Current research being undertaken in Namibia on ocean acidification, sea water harvesting, climate smart agriculture, and atmospheric science, is also highlighted in this chapter. The information presented in this chapter will be critical in guiding climate change mitigation policies in hyperarid African countries, thus reducing the burden caused by the global change in climate. Aspects on the direction of future research on climate adaptation with a holistic and multidisciplinary approach are also proposed.

Keywords

Climate smart agriculture · Aerosols · Shellfish · Marine environment · Clean energy · Seawater desalination

Introduction

Namibia is a very unique country within Southern Africa, being the most arid country in this region characterized by the existence of two deserts, namely the Namib and the Kalahari deserts. On average, Namibia receives 270 mm of

M. Haulofu
Water Quality Research, SANUMARC, Sam Nujoma Campus, University of Namibia, Henties Bay, Namibia
e-mail: mthomas@unam.na

S. K. Mafwila
Oceanography Research, SANUMARC, Sam Nujoma Campus, University of Namibia, Henties Bay, Namibia

Department of Fisheries and Aquatic Science, Sam Nujoma Campus, University of Namibia, Henties Bay, Namibia
e-mail: smafwila@unam.na

rainfall annually, with a wide variation of between 20 mm and 700 mm from the western to the eastern parts of the country, with only 5% of the country receiving more than 500 mm of rain annually (Sweet and Burke 2006). Within this hyperarid environment, only 1% of the total surface of the country has medium to high potential for rain-fed crop production. Interestingly, agriculture plays a critical role in the formal and informal economy of Namibia, supporting up to 70% of the population, with 90% of this agriculture being livestock based (MAWF 2015). Due to its hyperaridity, Namibia imports most of its food crop requirements from other countries. However, with the advent of climate change which threatens Africa's food production potential, hyperarid and food (especially grain and vegetable crops) importing countries like Namibia will require intensive research on how to mitigate against the threats paused by climate change.

Namibia has achieved some key milestones towards its response to climate change. Firstly, by being party to the United Nations Framework Convention on Climate Change (UNFCC), which meant it must institute policies and measures, informed by science, that reduce the changes on environment and humanity due to climate change (Mapaure 2011). In response to this, Namibia established the Namibian Climate Change Committee (NCCC) in 2001, to advice and make recommendations to government on issues of climate change. In 2011, Namibia developed the National Climate Change Policy (NCCP) which "outlines a coherent, transparent and inclusive framework on climate risk management in accordance with Namibia's national development agenda, legal framework, and in recognition of environmental constraints and vulnerability." (Lubinda 2015). In 2014, the country developed the National Climate Change Strategy and Action Plan for 2013–2020 which lays out guiding principles responsive to climate change that are effective, efficient, and practical (Lubinda 2015).

According to Mapaure (2011), Namibia has identified seven sectors where it is most vulnerable to climate change, and these include water resources; marine resources; agriculture; biodiversity and ecosystems; coastal zones and systems; and health and energy. Within the terrestrial agriculture environment, Namibia is most likely to experience more extreme weather patterns such as flooding and drought conditions under the changing climate. Furthermore, within the marine environment, it is predicted that the increase in northerly or easterly winds could result in suppressed upwelling in the nutrient-rich Benguela system, thus causing the accumulation of oxygen-poor water near the seabed and consequently suffocating marine life (Reid et al. 2008). Therefore, future changes in the distribution and intensity of winds under the changing climate have potential to negatively influence the fisheries sector of Namibia. Against this background, this chapter seeks to highlight some aspects critical for research on climate change in Namibia, while highlighting the current research being undertaken in Namibia at addressing climate change. This chapter focuses more on the coastal environment of Namibia, where climate change research on both terrestrial and marine environments is being championed at the Sam Nujoma Marine and Coastal Resources Research Center.

Namibia: A Vast Place

Namibia is located in the southwestern part of Africa between the latitudes of 17.5° and 29° South, occupying a total land area of about 824,000 km^2, with approximately 526,000 km^2 of ocean extending 200 nautical miles into the Atlantic Ocean (https://www.unicef.org/namibia). The country has a population of about 2.448 million people, making it one of the least populated country in the world. The country is characterized by a large coastal area, which covers the entire western border of the country with a length of about 1572 km. In Namibia, the cold Benguela Current has a moderating effect on regional weather patterns which are quite pronounced in the Namib desert, where it creates some unique climatic conditions. Just like most African countries, the contribution of Namibia to global atmospheric pollution responsible for climate change is very limited, though the effects of climate change will be quite pronounced. With Namibia's climate being already variable, climate change is expected to worsen this variability and amplify its adverse impacts (MET 2011). Therefore, with climate change, the focus for Namibia is more on mitigating its effects, which should be the focus of much research, as opposed to focusing on reducing greenhouse gas emissions as the country is not a serious contributor to this aspect.

Aerosol Monitoring in Namibia and Its Importance

Climate change poses great danger to the livelihoods of many people in Africa and Namibia is not spared from this phenomenon. The country recently experienced the worst drought in decades, that peaked from 2018 to 2019, with some areas receiving none to sporadic precipitation accompanied with floods. Since climate change is driven by several factors, many of these factors are a product of interactions in the earth's atmosphere. Due to these atmospheric interactions, it is important to do long-term baseline aerosol measurement in areas predicted to suffer most from effects of climate change. Though Namibia, like most African countries, does not contribute much to atmospheric pollution, aerosols from global pollution which include black carbon (soot), sulfate, nitrate, ammonium, organic carbon, elemental carbon, and mineral elements can play a critical role in Namibia's climate. Aerosols are suspended particles in the atmosphere, their sizes range from fine (particle diameter < 1 μm) to coarse (particle diameter > 1 μm). Aerosols play several roles of climatic importance as they influence the amount of precipitation by acting as condensation nuclei for cloud formation, while others depending on their nature may scatter and absorb solar radiation thus affecting radiative forcing (Heintzenberg 1985). Aerosol radiative forcing is defined as the effect of aerosols on the radiative fluxes at the top of the atmosphere and at the surface and on the absorption of radiation within the atmosphere (Chung et al. 2005).

The western coast of southern Africa, where Namibia is central, experiences persistent and extended highly reflective clouds mainly stratocumulus and high loads of light-absorbing aerosols from various sources (Formenti et al. 2018).

However, the mechanisms by which the reflective clouds and light-absorbing aerosols interact to influence the regional radiative budget is largely unknown (Formenti et al. 2018). In other words, the function of aerosols and cloud processes on climate models is still unclear. Due to the importance of different aerosols on regional climate, a ground-based aerosol monitoring station was established in Namibia known as the Henties Bay Aerosol Observatory (HBAO) in 2011 under the University of Namibia at their coastal campus (Sam Nujoma Campus) in collaboration with the Laboratoire Interuniversitaire des Systèmes Atmosphériques (LISA), CNRS/ Universities of Paris Est-Créteil and Paris Diderot, France, and the Climatology Research Group of the North West University (NWU), Potchefstroom, South Africa. This research station focuses on monitoring various parameters influenced by the aerosols. The data generated will form part of the atmospheric baseline studies in Namibia and the southern Africa region, which is important in filling the wide gap in climate change modelling in the region.

The contribution of greenhouse gases on climate change is well investigated, but due to the complex nature of aerosols, their contribution to climate still remains an area of interest to many scientists. As an arid country, Namibia's atmospheric precipitation could easily be affected by the nature of aerosols. If the amount of hydrophobic aerosols increases in Namibia, this could result in reduced rainfall and changes in the fog patterns in the country, which will negatively affect productivity of this arid but unique ecosystem.

In Namibia, the HBAO is monitoring the various properties of aerosols like those originating from the ocean (which generates microscopic salt particles by evaporation of water from sea-spray), wind-blown dust from the surrounding desert areas, dust from the various salt pans and aerosols generated by veld-fires, and smoke from ships using the Walvis Bay port. Aerosols from each of these sources will have a different composition, and different characteristics such as size, light's absorption capacity, and reflectivity.

In a 3-year study at the HBAO, measurement of mass concentration of equivalent black carbon (eBC, which is a light-absorbing aerosol from biomass burning), it was noted that these aerosols have seasonal shifts that coincide with fire seasons in southern Africa (Formenti et al. 2018), a clear indication that regional sources have contributed to the observations. In addition to these long-term measurements at the HBAO station, an intensive field campaign named the Aerosol Radiation and Clouds in southern Africa (AEROCLO-sA) project using ground-based equipment and a SAFIRE F20 research airplane was conducted in August to September 2017. The results from AEROCLO-sA field campaign are summarized by Formenti et al. (2019). Even though the sources of aerosols in Namibia are less studied, local and subcontinental sources may contribute to the observed measurements since finer aerosols are transported over a long distance. The southwestern African areas experience persistent stratocumulus cloud deck, which increases the total of outgoing radiation, resulting in a generally cooler atmosphere. Increased amount of eBC will therefore result in warmer atmosphere and may affect cloud formation in the region.

With the projection of increased emission in the world, Namibia is in a region that is predicted to experience rising temperatures and it is expected that a large portion of the country will see lowered rainfall with increased number of dry days consequently affecting agricultural productivity negatively. In addition, in desert and coastal areas the rate of fog production may be impacted resulting in less water for plants and animals in these areas. The overall effects of aerosols on climate can be complicated to predict, it is only a long-term observation that can assist in modeling and predicting the climate change regionally and globally.

Climate Change and the Oceans: Implications for Namibia

Rising atmospheric carbon dioxide (CO_2) concentrations became evident after the industrial revolution in the 1800s as a result of increased anthropogenic activities such as combustion of fossil fuels, agriculture, iron and steel production, municipal solid waste combustion, cement manufacturing, deforestation, and other activities that contribute to emissions of greenhouse gases. The industrial production of CO_2 since the 1800s has increased the atmospheric CO_2 concentration from 280 parts per million (ppm) to 396 ppm, and continued increases are predicted to reach 800 ppm by the year 2100 (Branch et al. 2013).

The current atmospheric CO_2 concentration is rising at an increased annual rate of 0.5%, that is 100 times faster than any change during the past 650,000 years. The global oceans absorb 25% of the CO_2 released into the atmosphere, from the various listed human activities. The increased absorption of CO_2 by the oceans has caused changes in the oceanic chemistry where the concentration of dissolved CO_2 has been increasing and consequently resulting in ocean pH decrease. This reduction in pH, as seawater becomes more acidic, is predicted to decline from approximately 8.2 during the preindustrial era to 7.8 in the 2100 s (Wei et al. 2014).

When CO_2 is absorbed by seawater, a series of chemical reactions occur resulting in the increased concentration of hydrogen ions which causes the seawater to become more acidic and causes carbonate ions to be relatively less abundant. The solubility and distribution of CO_2 in the oceans highly depends on climatic conditions and other physical (e.g., water column mixing, temperature), chemical (e.g., carbonate chemistry), and biological (biological productivity) aspects. Assimilation of CO_2 into the ocean occurs via the ocean surface to air interface, where $CO_2(aq)$ will react with surface seawater to form a weak carbonic acid (H_2CO_3) that quickly dissociates into hydrogen ions (H^+) and bicarbonate (HCO_3^-) ions, as indicated in the Eqs. (1) and (2) (Doney 2006).

$$CO_2 + H_2O \leftrightarrow H_2CO_3 \leftrightarrow HCO_3^- + H^+ \qquad (1)$$

Some of the free hydrogen ions (H^+) combine with free carbonate (CO_3^{2-}) ions normally found in seawater to form more bicarbonate (HCO_3^-).

$$H^+ + CO_3^{2-} \leftrightarrow HCO_3^- \qquad (2)$$

This chemical reaction uses up important ocean carbonate (CO_3^{2-}) molecules needed by some organisms, such as mussels and oysters, to form shells. In addition, the increased acidity of the ocean, caused by the free hydrogen ions (H^+), causes shells to deteriorate, become thin and frail. Shells in marine organisms are formed by taking up carbonate and calcium in seawater to form calcium carbonate, or the tough cement-like layer we know as a shell, as illustrated in Eq. (3).

$$CO_3^{2-} + Ca_2^+ \leftrightarrow CaCO_3 \tag{3}$$

Ocean acidification currently holds the world's attention due to the predicted adverse effects it poses on the marine ecosystems. Monitoring for ocean acidification is crucial in that it allows for forecasting future trends.

Ocean acidification is expected to impact marine species at varying degrees. While microalgae like dinoflagellates and diatoms may benefit from higher CO_2 conditions in the ocean, it has dramatic effects on calcifying species, such as oysters, clams, sea urchins, corals, and calcareous plankton. Increased sequestering of atmospheric CO_2 in the ocean would result in a lowering of the oceanic carbonate equilibrium. Carbonate ions are important building blocks for structures such as seashells and coral skeletons. Calcifying organisms such as corals are important in the ecosystem as they provide habitat and shelter to various marine organisms and their demise would endanger survival of most of those housed species. There are also other calcifiers that are vital food sources for most marine organisms. Since the calcification of seashells and coral reefs depends on the saturation of the carbonate mineral known as aragonite, it has been suggested that increasing atmospheric CO_2 could result in a decrease in the aragonite saturation state in the ocean by 30% (Takeo et al. 2012).

Namibia has a coastline that is largely uninhabited, unpolluted, and highly productive waters due to upwelling of the Benguela system. Upwelling systems are naturally acidic due to vertical movement of CO_2-rich deep waters to the surface and the acidity is increasing due to increased anthropogenic carbon dioxide that is being absorbed by the ocean. These upwelling systems are expected to undergo major changes in the near future due to steadily increasing atmospheric CO_2 and the resultant consequences on the ecosystem and fisheries are unknown. Mariculture in Namibia is reliant on shellfish culture with oyster farming being the most established activity, producing both the Pacific and Eastern oyster. Studies have shown how vulnerable mollusks are to acidification but there is still limited data on its impacts in Namibian waters (Waldron et al. 2009). Further research on the effects of ocean acidification on shellfish is being undertaken at the Sam Nujoma Marine and Coastal Resources Research Center, which promises better understanding of this phenomenon on marine life.

The Pacific oysters are farmed in Swakopmund, Walvis Bay, and Lüderitz (now known as !Nami‡Nûs). This industry is already under threat from the harmful algal blooms (HABs) that frequently occur along the Namibian coastline attributed by high plankton biomass decay that ultimately leads to anoxic events and in some cases eruptions of hydrogen sulfide from the seabed that led to 80% mortality of marine shellfish farmed in Namibia (Anderson et al. 2004). In 2008, sulfur eruptions impacted Walvis Bay marine shellfish farms leading to closure of some farms and relocation of other farms to Lüderitz (OLRAC 2009). With the Namibian marine industry already being influenced by the HAB, ocean acidification of the Southern Atlantic poses a far much greater threat to this industry.

Villafañe et al. (2015) indicated that the predicated ocean acidification conditions, with increased nutrient and solar radiation could lead to the reshaping of the phytoplankton community, where it would have phytoplankton communities dominated by large diatoms with high growth rates and therefore marine phytoplankton could bloom more efficiently under future acidified conditions than the current conditions. Similarly, when conditions favor phytoplankton, it also infers good conditions for growing seaweed. This would also suggest that the Namibian mariculture industry would lose high-income businesses from oyster culture, loss of employment, and subsequently reduced contribution to the country's GDP. Most businesses would have to shift into seaweed culturing as this will utilize the abundant nutrients, solar radiation, and CO_2 in the Benguela upwelling system. Therefore, research that already looks at macro-algae culture for extraction of highly priced phytochemicals is important in Namibia.

Previous research has found out that influxes of iron from terrestrial origin into the ocean led to increased bloom activity in primary productivity (Pollard et al. 2009). Research that involves fertilizing certain oceanic regions with iron, as it is mostly a limiting nutrient, to enhance the sequestering of carbon onto the ocean floor can be critical in mitigating against increased CO_2 concentration in the ocean. The resulting blooms would counteract the growing concern of dissolved CO_2 and produce organic carbon that would sink to the ocean floor. Norton (2011) suggested the addition of powdered limestone to the ocean in order to react with the dissolved CO_2 as another potential mitigation to ocean acidification. As the limestone reacts with the CO_2 this will form bicarbonate which would in turn neutralize the acidic conditions of the ocean. These counter action would make calcium carbonate more available in seawater and readily available to calcifying organisms to build their shells. The process of increased calcium carbonate would increase the buffering capacity of the ocean and limit the sequestration of CO_2 by the oceans. However, though these researched measures are theoretically feasible, their practicality still requires lot of practical field-based research.

Alternative Water Sources in a Hyperarid Namibia

Availability of Water in Namibia

Due to pressures from climate change, global water resources have become vulnerable and scarce, and Namibia is not spared from this phenomenon. Water is a significant resource to human life, the environment and for development, but it is finite. Scientific consensus is that climate change will have a pervasive influence on the future demand, supply, and quality of fresh water resources in Namibia, and would add pressure to already stressed water resources and coastal environments (Lahnsteiner and Lempert 2007). Namibia frequently receives below average and highly variable rainfall with high temperatures that result in very high evaporation rates of between 2660 and 3800 mm per year, which makes Namibia one of the driest in sub-Saharan Africa (FAO 2005). More specifically, the central Namib Desert consists of a flat, gently sloping plain with few topographic features. The steady gradients affect rainfall, temperature, and wind patterns developing between the coast and the interior. The annual mean rainfall along the coast between Swakopmund and Henties Bay is less than 15 mm per annum, with coastal wind speeds averaging approximately 3 m/s in a south westerly direction (Pryor et al. 2009). Likewise, due to climate change, the occurrence of droughts has become more frequent and requires appropriate water management to ensure continued social and economic development of the country (COW 2019). Due to extremely poor rainfalls, the recent drought of the 2018/2019 rainfall season resulted in huge economic loses to the Namibian farmers as livestock and crops died due to lack of water.

The rainfall received accounts for only 2% and 1% surface runoff and groundwater recharge, respectively, while 97% is lost through high evaporation rates in Namibia (Christelis and Struckmeier 2011). The difference between national and regional averages reveals the growing need to seek and implement alternative water supply strategies, in order to secure adequate fresh water and cater for future demand in Namibia under a changing climate. Namibia relies heavily on underground water (aquifers) for the supply of fresh water, which are recharged not only by the rainfall from catchment areas within the country but also from neighboring countries. With such heavy reliance on underground water, particularly in the North Central Regions, the water is at times unsuitable for human and animal consumption due to high salinity (Wilhelm 2012). Climate change is expected to exacerbate this reliance on aquifers for fresh water due to limited recharging of underground water sources. Changes in rainfall and runoff further indicate that underground water recharge may suffer a reduction of 30–70% across Namibia. In order to adapt to the reduction in freshwater resources, other arid countries including Namibia have begun sea water desalination as a feasible source of fresh water. The reliance is expected to grow as a result of population growth and migration to coastal cities which do not have adequate freshwater resources. In Namibia, the coastal population's overdependence on aquifers constitutes its vulnerability to climate change. However, the vast coastal area of Namibia is also characterized by cool yet sunny environments, rough seas, and very windy areas, all which can be potential energy sources for powering the energy demanding and costly desalination process. Desalination would be a strategic alternative to ensure long-term freshwater supplies, more so when using renewable, clean energy sources for desalination.

Seawater Desalination in Namibia

On earth, about 70% of the area is covered by water which is mainly seawater, with only about 1% of the freshwater being available for human consumption and for agricultural purposes in the world's rivers, lakes, and aquifers (Darre and Toor 2018). The scarcity of fresh water resources and the need for additional fresh water supplies are critical in many arid regions of the world (Michou 2017). Worthy to note are the countries in the Arabian Gulf, which are located in an arid area with limited water resources. The region has the lowest availability of renewable water resources per capita in the world. To curb this shortage, the gulf countries invested in seawater desalination technologies that would ensure an adequate supply of fresh-water for industries and the more than 300 million inhabitants. To date, the largest market for water desalination in the world is found in the Arabic Gulf (Ramadan 2015).

Even though it offers a sustainable solution to water woes, desalination is energy intensive and comes with environmental consequences. About 0.4% of the world's electricity consumption is taken up by desalination plants, with global emissions due to nonrenewable energy use in desalination, predicted to reach 400 million carbon equivalents per year. The carbon footprint for reverse osmosis (RO) desalination has been calculated to be between 0.4 and 6.7 kg CO_2eq/m^3. This means that desalinating 1000 cubic meters of seawater could potentially release as much as 6.7 tons of CO_2 (Tal 2018), hence efforts to produce freshwater should not forgo environmental safety. Desalination technologies must be enhanced to become cleaner and more efficient (Compain 2012).

Desalination needs are mainly in dry countries that receive high-intensity solar radiation, hence it makes sense in Namibia to exploit solar power for desalination. Namibia is a country which is abundantly rich in solar energy, as it receives on average of 300 days of sunshine per year, with very few cloudy days. Harnessing this energy from the sun would make it possible to supply the power needed to run desalination plants. The available solar energy makes research and development in solar desalination promising for Namibia. Although at present, renewable resources in Namibia play a very small role in the energy sector, there is potential for renewables to scale up in providing for non-electricity energy sectors. Other alternative and cost-effective sources of renewable energy can also be used for desalination. Renewable energy sources such as wind can be used to generate electricity for use in desalination plants as an alternative to for thermal electric plants. The above alternatives are already being explored in the east coast of Egypt where water resources are limited.

In Namibia, the first desalination plant belonging to Orano Mining Namibia was commissioned in 2010. The desalination plant was originally built to supply water to the Trekkopje Mine near Arandis but has now become part of the potable water delivery system belonging to Namibia's water utility, NamWater. The addition of desalinated water has allowed NamWater to keep meeting the Central Namib's water demand as it provides about 75% of the overall drinking water to the coastal town of Swakopmund, as well as the nearby uranium mines and other industries (Petrick and Müller 2019). Furthermore, communities of Akutsima and Amarika in the Omusati Region of Namibia benefited from small-scale solar powered desalination plants under the CuveWaters project. These plants aimed to supply the local population with safe drinking water by desalinating saline underground water. This was seen as a sustainable way of supplying safe drinking water to the local community, who for years consumed unsafe and salty water.

In an effort to reduce the carbon footprint of desalination, the University of Namibia piloted a 100% solar powered desalination plant that uses reverse osmosis technology. The pilot desalination plant is evaluating the feasibility of using renewable solar energy for seawater desalination in Namibia's coast. This pilot research project generates approximately 2300 L of fresh water per hour, which can be very sustainable for supplying water to local communities. Preliminary results of an assessment to determine the mineral composition and grade class of the desalinated water produced at the University of Namibia (UNAM) are presented in Table 1, which point to the suitability of this water for drinking and irrigation according to National and WHO standards.

With this research solar powered plant, the University of Namibia seeks to evaluate various scenarios which include:

- The short- and long-term suitability of solar power in water desalination in the Namibian coast.
- The potential of using this clean energy produced fresh water in desert greening and food production.
- The potential of using desalinated water in enhancing Namibia's contribution to carbon sequestration by using the water to irrigate high carbon fixing trees.
- The potential of accruing carbon credits using various indigenous and exotic tree species irrigated using desalinated water in Namibia.

Table 1 Preliminary results on the mineral composition and grade class of UNAM desalinated seawater

Physical constituents	UNAM	National limits	WHO limits	Units	Grade	Classification
pH	5.9	6–8.5	6.5–8.5		B	Good
Turbidity	0.3	<0.3	< 4	(NTU)	A	Excellent
Conductivity	45.2	<80	[a]	(mS/m)	A	Excellent
Total hardness as CaCO$_3$ (mg/L)	7	<400	[a]	mg/L	A	Excellent
Mineral constituents						
Sodium as Na (mg/L)	55	<100	200	mg/L	A	Excellent
Potassium as K	3	< 25	[a]	mg/L	A	Excellent
Nitrate as N (mg/L)	<0.5	< 6	< 50	mg/L	A	Excellent
Nitrite as N	<0.1	< 0.1	< 3	mg/L	A	Excellent
Fluoride as F (mg/L)	<0.1	< 0.07	1.5	mg/L	A	Excellent
Iron as Fe (mg/L)	0.01	< 200	<0.3	mg/L	A	Excellent
Calcium as CaCO$_3$ (mg/L)	<3	< 80	200	mg/L	A	Excellent

[a]Occurs in drinking water at concentrations well below those of health concerns (WHO 2009)

Alternative Food Sources under a Changing Climate: Mushrooms

Though Namibia is an arid country, with very limited terrestrial productivity, there has been a growing interest on mushrooms (edible macrofungi) as an alternative high-value food under the changing climatic environment. Edible macrofungi are highly nutritious with numerous therapeutic benefits as they are rich in proteins, minerals, and vitamins such as B, C, and D (Reeta and Dev 2019). In Namibia, foraging for edible macrofungi is done seasonally just like elsewhere in sub-Saharan Africa (Wahab et al. 2019). Edible macrofungi are mostly foraged for personal consumption but some individuals sell them for a small price, pointing to their potential economic value. Interestingly, those who sell them are unwilling to profit from their sale as they believe that macrofungi are God's given gift (Mshigeni 2001).

Just like other wild crops in Namibia, the availability of macrofungi is influenced by rainfall, with too little or too much having a negative effect on their distribution and abundance. Ethno-mycological studies have reported that Namibian people are aware of the link between the appearance of macrofungi with rainfall, since they refer to the macrofungi as products of a "lightning bird," because they appear after thunderstorms (Trappe et al. 2010). Of recent, wild edible macrofungi are diminishing with unpredictable rainfall events. The unusually heavy rains within a short period, followed by extended drought periods, is reducing the natural growing season of mushrooms. There is therefore a need to shift from the traditional way of exploiting this resource under a changing climate by focusing on cultivation technology. Mushroom cultivation has potential to contribute to diversification of traditional farming in Namibia. Despite the global trend in domestication and cultivation of macrofungi, the focus in Namibia to date has been on the cultivation of the exotic oyster mushrooms (Hausiku and Mupambwa 2018). Even with the oyster mushroom, their cultivation is still on a subsistence scale, partly due to shortages of suitable substrate. As for the indigenous macrofungi, a limited number of research has been done on characterizing their chemical properties and their domestication studies are still in infancy stage if there are any.

Climate Change Implications on Cultivation of Macrofungi in Namibia

Since macrofungi thrive underground or within substrate during their vegetative stage, their direct response to climate change is challenging to characterize. Consequentially, the timing of their fruit body production is used to study their phenological response to climate change (Bidartondo et al. 2018). Sato et al. (2012) defined phenology as the seasonal timing of life-history events of organisms (in this case, macrofungi). For the Namibian macrofungi, the only documented phenological response to climate change is that of the *Kalaharituber,* also known as Kalahari truffle (Mshigeni 2001; Leistner 1967) showing a later and shorter fruiting period. This observation is in agreement with long-term datasets documented elsewhere showing that the lengths of the macrofungi-producing season and their timing have changed relative to the changes in temperature and rainfall, as have their diversity and range (Bidartondo et al. 2018). It can therefore be deduced that the rising temperature and reduction in precipitation observed in Namibia over the years is having similar effects on macrofungi productivity in Namibia. Studies carried out in the laboratory confirm that temperature and moisture availability affect both the vegetative and fruiting stage of macrofungi. Mycelial growth has been reported to increase with temperature until an optimum temperature is reached, thereafter, the growth decreases (Fletcher 2019). Similar response has been observed in the presence of moisture, whereby insufficient or excessive moisture limits fungal growth. According to Wahab et al. (2019), the average optimal temperature and humidity for mushroom fruiting is around 20 °C and 80%, respectively.

Current and Future Research on Mushrooms in Namibia

The projected further increase in temperature and sporadic rainfall patterns coupled with extreme weather events are likely to threaten the already dwindling macro-fungal resources of socioeconomic value in Namibia. A few studies have been documented on Namibian macrofungi as summarized in Table 2, and a lot more need to be done in terms of research.

From Table 2, it is obvious that Namibian macrofungi are understudied. All the documented studies on the edible macrofungi (*Termitomyces* and *Kalaharituber* species) were done on specimens that were bought from the market. Further studies are needed to address the distribution, ecology, phylogeny, pharmacology, phenology, chemical characterization, and domestication trials.

More studies are documented on nonedible macrofungi (*Trametes*, *Pycnoporu, and Ganoderma* species) compared to the edible ones. These studies addressed the phylogeny, indigenous knowledge, and the utilization of these fungi. Further studies are needed to zoom into the biochemical properties of the fungi to isolate and quantify the bioactive compounds responsible for their therapeutic effects reported by the respondents during these studies. Phenological studies would give an idea on the effect of climate change on the different species. Domestication trials would facilitate intensification of research on the fungi as field studies tend to be time-consuming and expensive.

Namibian mycologists are thus urged to make an effort to domesticate indigenous macrofungal species, especially the edible ones that have largely been neglected by science. Domestication and cultivation will ensure that availability of this resource is not dictated by precipitation or climate. It has been demonstrated by research that cultivation of macrofungi depends on agro-industrial material (e.g., cereal straws, wood chips, etc.) for substrate, and these materials are usually rain fed implying that drought effects on substrate material will trickle down to mushroom productivity. Recent research has been exploring the potential of using marine biomass as a source of organic material to be used as mushroom substrate. In a study by Hausiku and Mupambwa (2018), beach-casted seaweed was used to supplement rice straw in order to fruit oyster mushrooms. The incorporation of seaweed into substrate did not

Table 2 A summary of studies that have been done on indigenous macrofungi in Namibia

Macrofungal species	Type of study	Guide for future studies
Termitomyces schimperi and *Kalaharituber pfeilii*	Antimalarial properties of *Termitomyces schimperi* and *Kalaharituber pfeilii*	Distribution, ecology, phylogeny, pharmacology, phenology, chemical characterization, antimicrobial activities, domestication trials.
	Characterization of polysaccharides isolated from *Termitomyces schimperi* and *Kalaharituber pfeilii*	
	Characterization of laccase enzymes from *Termitomyces schimperi* and *Kalaharituber pfeilii*	
	Antiplasmodial activity of *Terfezia pfeilii*	
	Morphological characterization of *Termitomyces schimperi* from herbarium specimens	
General macrofungi	Survey of Namibian macrofungi	Spatial and temporal distribution and habitats of the different species, phylogeny, phenology over time
Trametes and *Pycnoporus*	Phylogeny of *Trametes* and *Pycnoporus* using ITS	Biochemical characterization, phenology, domestication trials
	Ethnomycology of indigenous *Trametes* species	
Ganoderma lucidum	Antimalarial properties of *Ganoderma lucidum*	Distribution of the fungus over time. Characterization and quantification of bioactive compounds
	Minerals and trace elements in *Ganoderma lucidum*	
	Evaluation of substrate for the production of *Ganoderma lucidum*	
	Antimicrobial screening of crude extracts from *Ganoderma lucidum*	
	Antiplasmodial activity *Ganoderma lucidum*	
	Traditional medicinal uses and natural hosts	
	Genetic diversity of *Ganoderma*	
	Distribution, genetic diversity, and uses of Ganoderma	

only add value to the beach-casted seaweed but enhanced the nutritional benefits of the mushrooms.

Identification, collection, and domestication of native wild mushrooms is another aspect of fungal science that have been overlooked. Such studies are imperative in facilitating taxonomical and ethno-mycological studies of indigenous macrofungi. Furthermore, documentation on the traditional use of Namibian macrofungi and assessment of their local and international markets would expedite their industrial relevance (Shahtahmasebi et al. 2018). Besides domestication, value addition of macrofungi with socioeconomic importance such as the *Termitomyces* and the *Kalaharituber* is important. Food technologists need to find ways to increase the shelf life of this highly perishable commodity by creating new products that are different from the traditional forms in which macrofungi are consumed in order to contribute to the country's GDP.

Organic Agriculture and Its Potential Under a Changing Climate

With Namibia being an arid country, terrestrial biomass production is very limited, which has consequently resulted in very poor quality soils in most parts of the country. The soils in Namibia therefore lack fertility as they are mostly sandy and very old (Zimmermann et al. 2017). Under natural conditions, terrestrial biomass results in the generation of huge quantities of organic matter which is deposited into the soil, where it becomes part of soil organic carbon. Soil organic carbon or soil organic matter plays a critical role in soil quality, with the following functions:

- The non-humic substances produced during the decomposition of organic matter, e.g., polysaccharides encourage granulation – the binding of soil particles into aggregates. These aggregates help to maintain a loose, open, granular condition. This improves the aeration and water infiltration capacity of the soil.
- Organic matter can hold up to 20 times its weight in water. Its presence in the soil, therefore, improves the soils' water holding capacity. This is especially important in sandy soils.
- Organic matter (humus colloids) has a high cation exchange capacity (2–30 times as great (per kg) as that of various types of clay minerals). It therefore adsorbs and stores plant nutrients (K^+ Ca^{2+}, Mg^{2+}, etc.) and thus improves the fertility of soil.
- Humus colloids form stable complexes with Cn^{2+}, Zn^{2+}, Mn^{2+}, and other poly-valent cations. This may enhance the availability of micronutrients to higher plants.
- Decomposition of soil organic matter release NH_4+, NO_3^-, PO_4^{3-}, and thus serves as a source of nutrients to plants.
- Organic matter serves as a source of energy for both macro- and microfaunal organisms. These organisms in turn play important roles in soils, e.g., earthworms which are important agents for good soil structure, bacteria responsible for mineralization processes, etc.
- Humic acids also attack soil minerals and accelerate their decomposition, thereby releasing essential elements as exchangeable cations.
- In very acid soils, organic matter alleviates aluminum toxicity by binding Al^{3+} ions in nontoxic complexes

Soils with low organic matter are therefore prone to degradation and this is the major cause of reduced agricultural productivity and food security in countries like Namibia. With poor quality soils mainly due to low organic matter content, much of Namibian agriculture is more at risk of the effects of climate change such as reduced rainfall, flooding, increased environmental temperature among others, which will require more resilient soils.

From the Namibian perspective, a shift from the conventional agriculture that replicates the green revolution technologies such as intensive use of fertilizers, pesticides, and machinery is required, towards more organic-based agriculture systems is important. The current conventional agriculture practices have not focused on soil fertility but rather only on the crop fertility, which has actually seen crops yields in most parts of the country decreasing in the face of climate change. Organic agriculture practices are among some of the systems that are being promoted among others like ecological agriculture, sustainable intensification, and conservation agriculture, as these systems focus more on feeding the soil making it more resilient under climate change (Mupambwa et al. 2020). Organic agriculture is a more holistic production system which promotes agroecosystem health approach which promotes the use of management practices in preference to the use of off-farm inputs (FAO/WHO 1999).

Zimmermann et al. (2017) highlighted that Namibia has important resources that can be used in organic soil fertility which include animal manures, fertilizer trees harvestable from nature; marine resources such as seaweeds; bird guano and animal processing waste such as blood, bones, and fish waste. Technologies like composting and vermicomposting can be used in improving the fertilizer value of these waste materials, which can then be used as soil amendments. Current research being undertaken in Namibia on developing organic nutrient-rich fertilizers includes the optimization of the biodegradation and nutrient release during vermicomposting of various household waste, marine biomass, and farm waste materials. Such research is important in standardizing the production processes of organic fertilizers, thus allowing for greater uptake of these fertilizers among smallholder farmers. The lack of nutrient predictability and standardization of organic fertilizer production results in even those that are commercially available on the market being sold without any labeling that indicates nutritional composition as is mandatory for inorganic fertilizers. The specification of compost quality is more important as it enables compost users to more effectively manage their agroecosystems (Mupambwa and Mnkeni 2018).

Several researchers are undertaking research that is linked to organic agriculture in Namibia and these are summarized in Table 3 below.

Research on uses of organic soil fertility conditioners is currently happening in Namibia, though not much of it is not directly linked to mitigating the effects of climate change. There is a need to marry the two, so that such research can directly fit into the National Climate Change Strategy and Action Plan for 2013–2020 of Namibia. It is also important to note that there is very limited research available for Namibia on evaluating the potential of organic agriculture in improving soil quality, though several reviews are available on this subject.

Table 3 A summary of some of the studies that have focused on aspects relating to organic agriculture in Namibia

Study details and location	Key results	Climate mitigation effects	Future research	Reference
The study evaluated the used of three rates of cattle manure at 0, 31, and 62 Mg/ha within three cropping systems, i.e., pearl millet (*Pennisetum glaucum* L.) mono-cropping; cowpea (*Vigna unguiculata* L.) mono-cropping; and pearl millet-cowpea intercropping. The soil was a sandy soil from Ogongo, Omusati Region with a pH (H_2O) = 7.0 and 2.8 g total C/kg soil.	Application of manure increased soil organic carbon by 1.3 and 1.7 times under the 31 and 62 Mg/ha, respectively. Furthermore, manure application significantly increased the concentration of most macronutrients compared to the control. However, cropping system did not result in differences in most of these parameters. The study also observed that there was optimization on the benefits of manure addition at the lower rate of 31 Mg/ha.	Soil organic carbon (SOC) can be significantly increased by addition of cow manure, which has the potential to improve soil quality in sandy degraded soils. Higher SOC increases the resilience of soils and improves water and nutrient retention, which has potential to increase yields under smallholder farmers.	There is need to improve the quality of the manure applied through processes such as composting and vermicomposting and evaluating these improved organic nutrient sources. There is need for long-term studies evaluating the influence of animal manures on soil quality, which considers changes soil chemical, biological, and physical properties. Optimization of long-term application rates of animal manures which is informed by crop requirements is essential.	Watanabe et al. (2019)
The study evaluated an organic hydroponic nutrient solution prepared from goat manure processed through composting goat manure for 6 months and then extracting the nutrients from the compost by a process of leaching for 7 days. The study	The use of organic hydroponic solution did not result in significant differences in vegetative plant growth with the commercial hydroponic fertilizer, which were all different from the control with the lowest vegetative growth In terms of yield parameters,	Important vegetable crops like tomatoes can be grown in under hydroponic systems that are fertilized using organic nutrient sources. The use of hydroponics which are cheap allows for the production of crops under the smallholder farmers with limited water use	Evaluating more nutrient-rich compost types called vermicomposts as hydroponic nutrient sources. Evaluating the effectiveness of using amendments like rock phosphate, fly ash, and biochar in improving the fertilizer value of these organic nutrient sources.	Mowa (2015) and Mowa et al. (2017)

(continued)

Table 3 (continued)

Study details and location	Key results	Climate mitigation effects	Future research	Reference
had three treatments, i.e., the organic hydroponic fertilizer; a commercial hydroponic fertilizer, and a control of water alone. A floating deep water culture hydroponic system was then used for growing tomatoes. The study was done in a greenhouse in Henties Bay located within the coast of Namibia.	tomato plants grown under the commercial hydroponic fertilizer produced more than those grown under manure nutrient solution which in turn, produced more than the plants grown under the control.	and low fertilizer costs. Hydroponics are critical in increasing food production under a changing climate, where fresh water will become scare.		
The study was undertaken in Kavango East Region (Mashare Irrigation Training Center – MITC) and Zambezi Region (Liselo Research Station – LRS). Two experiments one with pearl millet (*Pennisetum glaucum*) and the other with maize (*Zea mays*); which were rotated with different Green Manure Cover Crops (GMCC) to evaluate their effects on the productivity (grain and biomass yields) among other parameters.	The GMCC lablab resulted in the highest biomass yield with a mean yield of 11,500 kg/ha while jack bean had the lowest yield of 1000 kg/ha. Similarly, lablab in the pearl millet rotation resulted in the greatest grain yield of 3 tonnes/ha, with jack bean, bambara nut, and common sun hemp producing lower grain yields. At LRS, the greatest biomass yield was observed in the velvet bean treatment with a mean yield of 5900 kg/ha.	Green manure cover crops can significantly contribute to the addition of soil organic matter thus increasing the soil organic carbon of the soil, which consequently improves soil quality. The legume green manure cover crops can be effective in fixing atmospheric N into the soil, thus reducing the use of inorganic fertilizers.	Monitoring the long-term contribution of the GMCC to overall soil quality under various combinations. Evaluating the potential of agroforestry tree species in agroecology in Namibia.	Amakali (2019)

The contribution of Namibia and most countries in southern Africa to climate change through the global increase in atmospheric pollutants that are drivers of global warming is very minimal. However, it is apparent that the impacts of this global change in climate due to human-induced activities will influence everyone on earth, with the effects expected to be even more pronounced in already hyperarid countries like Namibia. Namibia has made huge progress at policy level with regards to issues of climate change. However, there is need for more urgent research that directly feeds into the climate change strategy and action plan for Namibia. This research will need to focus on understanding more on the effects of anthropogenic activities on the atmosphere, water, and land activities, in a holistic manner. With the country being hyperarid, research that is currently underway on using clean energy sources like solar, wind, and tidal power for sea water desalination can be critical in abetting the effects on climate change on fresh water availability in Namibia. However, there is need for research on engineering locally adaptable technologies that can take this fresh water from the coast to other places of the country for use in agricultural activities, thus creating green belts in Namibia, capable of improving the country's food self-sufficiency. With Namibia being blessed with a vast marine environment, research on the effects of climate change on this unique marine environment and how it will affect economic activities in this industry will also be critical. Overly, Namibia is making progress on generating research critical in mitigating against climate change, and the future will require more research with a multidisciplinary approach to feed into the country's climate change action plan.

References

Amakali S (2019) Effects of green manure cover crops on weed population, pearl millet and maize productivity under conservation agriculture in Liselo and Mashare Namibia. MSc thesis, University of Namibia

Anderson P, Currie B, Louw DC, Anderson DM, Fernández-Tejedor M, McMahon T, Rangel I, Ellitson P, Torres O (2004) Feasibility study for cost-effective monitoring in Namibia and Angola with an analysis of the various options for implementation of Shellfish Safety Programs. Final Report. BCLME project EV HAB/02/02a: Development of an Operational Capacity for Monitoring of Harmful Algal Blooms in Countries Bordering the Northern part of the Benguela Current Large Marine Ecosystem Phase 1 –Design

Bidartondo MI, Ellis C, Kauserud H, Kennedy PG, Lilleskov EA, Suz LM, Andrew C (2018) Climate change: fungal responses and effects. In: Willis KJ (ed) State of the world's fungi. Report. Royal Botanic Gardens, Kew, pp 62–69

Branch TA, DeJoseph BM, Ray LJ, Wagner CA (2013) Review – impacts of ocean acidification on marine seafood. Trends Ecol Evol 28(3):178–186

Christelis G, Struckmeier WF (2011) Groundwater in Namibia. In Water

Chung CE, Ramanathan V, Kim D, Podgorny IA (2005) Global anthropogenic aerosol direct forcing derived from satellite and ground-based observations. J Geophys Res 110:D24207. https://doi.org/10.1029/2005JD006356

Compain P (2012) Solar Energy for Water desalination 46(0):220–227. https://doi.org/10.1016/j.proeng.2012.09.468

COW (2019) Water management plan. City of Windhoek, Windhoek

Darre NC, Toor GS (2018) Desalination of water: a review desalination of water: a review. March. https://doi.org/10.1007/s40726-018-0085-9

Doney SC (2006) The dangers of ocean acidification. Sci Am 294(3):58–65

FAO (2005) Namibia Aquastat Survey. 1–14. http://www.fao.org/nr/water/aquastat/countries_regions/NAM/NAM-CP_eng.pdf

FAO/WHO (1999) Codex Alimentarius commission approves guidelines for organic food. http://www.fao.org/organicag/oa-faq/oa-faq1/en/. Accessed 23 June 2020

Fletcher IA (2019) Effect of temperature and growth media on mycelium growth of *Pleurotus ostreatus and Ganoderma lucidum* Strains. Cohesive J Microbiol Infect Dis 2(5). ISSN 2578-0190. https://doi.org/10.31031/CJMI.2019.02.000549

Formenti P, John Piketh S, Namwoonde A, Klopper D, Burger R, Cazaunau M et al (2018) Three years of measurements of light-absorbing aerosols over coastal Namibia: seasonality, origin, and transport. Atmos Chem Phys 18(23). https://doi.org/10.5194/acp-18-17003-2018

Formenti P, D'Anna B, Flamant C, Mallet M, Piketh SJ, Schepanski K, ... Holben B (2019) The aerosols, radiation and clouds in southern Africa field campaign in Namibia overview, illustrative observations, and way forward. Bull Am Meteorol Soc, 100(7):1277–1298. https://doi.org/10.1175/BAMS-D-17-0278.1

Hausiku MK, Mupambwa HA (2018) Seaweed amended rice straw substrate and its influence on health related nutrients, trace elements, growth and yield of edible white elm mushroom (Hypsizygus ulmarius). Int J Agric Biol 20:2763–2769

Heintzenberg J (1985) What can we learn from aerosol measurements at baseline stations? J Atmos Chem 3(1):153–169. https://doi.org/10.1007/BF00049374

Lahnsteiner J, Lempert G (2007) Water management in Windhoek, Namibia. Water Sci Technol 55 (1–2):441–448. https://doi.org/10.2166/wst.2007.022

Leistner OA (1967) The plant ecology of the southern Kalahari, Republic of South Africa. Bot Res Inst Bot Surv Mem 38:1–172

Lubinda M (2015) Factsheet on climate change: the definition, causes, effects and responses in Namibia. https://www.thinknamibia.org.na/files/learn-and-engage/TlaQWEcwS5WDlWme.pdf. Accessed 22 June 2020

Mapaure I (2011) Climate change in Namibia: projected trends and effects. In: Ruppel OC, Ruppel-Schlichting K (eds) Environmental Law and Policy in Namibia. Hans Seidel Foundation, Windhoek, pp 291–309

MET (Ministry of Environment and Tourism) (2011) National policy on climate change for Namibia. Government of the Republic of Namibia, Windhoek

Michou T (2017) Water and greenhouse gases: the desalination challenge

Ministry of Agriculture Water and Forestry (MAWF) (2015) Namibia agriculture policy. Ministry of Agriculture Water and Forestry, Windhoek

Mowa E (2015) Organic manure for vegetable production under hydroponic conditions in arid Namibia. Int Sci Technol J Namibia 5:3–12

Mowa E, Akundabweni L, Chimwamurombe P, Oku E, Mupambwa HA (2017) The influence of organic manure formulated from goat manure on growth and yield of tomato (Lycopersicum esculentum). Afr J Agric Res 12(41):3061–3067. https://doi.org/10.5897/AJAR2017.12657

Mshigeni KE (2001) The cost of scientific and technological ignorance with special reference to Africa's rich biodiversity. University of Namibia, Windhoek. https://doi.org/10.4314/dai.v13i3.15605

Mupambwa HA, Mnkeni PNS (2018) Optimizing the vermicomposting of organic waste amended with inorganic materials for production of nutrient-rich organic fertilizers: a review. Environ Sci Pollut R 25:10577–10595

Mupambwa HA, Ravindran B, Dube E, Lukashe NS, Katakula AAN, Mnkeni PNS (2020) Some perspectives on Vermicompost utilization in organic agriculture. In: Bhat SA, Vig AP, Li F, Ravindran B (eds) Earthworm assisted remediation of effluents and wastes. Springer, Japan, pp 299–331

Norton JM (2011) Ocean acidification: cause, effect, and potential mitigation approaches. A Journal of Academic Writing, Hohonu, p 9

OLRAC (2009) Specialist study: marine ecology impact assessment. Environmental impact assessment for the proposed expansion of the container terminal at the port of Walvis Bay. Prepared by OLRAC (Ocean and land resource assessment consultants) for CSIR Stellenbosch

Petrick W, Müller S (2019) Erongo desalination plant environmental management plan, Swakopmund

Pollard RT, Salter I, Sanders RJ, Lucas MI, Moore MC, Mills RA, Statham PJ, Allen JT, Baker AR, Bakker DCE, Charette MA, Fielding S, Fones GR, French M, Hickman AE, Holland RJ, Hughes Aj, Jickells TD, Lampitt RS, Morris PJ, Nedelec FH, Nielsdottir M, Planquette H, Popova EE, Poulton AJ, Read JF, Seevave S, Smith T, Stinchcombe M, Taylor S, Thomalla S, Venables HJ, Williamson R, Zubkov MV (2009) Southern Ocean deep-water carbon export enhanced by natural iron fertilization. Nature 457:577–580

Pryor AM, Blanco B, Galtes J (2009) Desalination and energy efficiency for a uranium mine in Namibia. Ground Water:1–16

Ramadan E (2015) Sustainable water resources management in arid environment: the case of Arabian Gulf. Int J Waste Resour 05(03):3–6. https://doi.org/10.4172/2252-5211.1000179

Reeta M, Dev MY (2019) Performance of oyster mushroom through use of different agro Byproducts. J Community Mobilization Sustain Dev 14(1):174–178

Reid H, Sahlen L, Stage J, Macgregor J (2008) Climate change impacts on Namibia's natural resources and economy. Clim Pol 8:452–466

Sato H, Morimoto S, Hattori T (2012) A thirty-year survey reveals that ecosystem function of fungi predicts phenology of mushroom fruiting. PloS one 7(11):e49777. https://doi.org/10.1371/journal.pone.0049777

Shahtahmasebi SH, Pourianfar HR, Mahdizadeh V, Shahzadeh Fazeli SA, Amoozegar MA, Nasr SH, Zabihi SS, Rezaeian SH, Malekzadeh KH, Janpoor J (2018) A preliminary study on domestication of wild-growing medicinal mushrooms collected from Northern Iran. Journal of Fungal Biology 8(6):606–623

Sweet J, Burke A (2006) Country pasture/forage resource profiles Namibia. In: FAO. Namibia climate-smart agriculture program. Food and Agriculture Organization of the United Nations (FAO), Windhoek, pp 2015–2025

Takeo H, Shoko K, Koichi S, Yuhi S, Yuko O, Shigeki W, Taiki A, Shun H, Takashi M, Masao I, Shu S, Daisuke S, Hiroko E, Tsuyoshi N, Isao I (2012) Effect of ocean acidification on coastal phytoplankton composition and accompanying organic nitrogen production. J Oceanogr 68:183–194

Tal A (2018) Addressing desalination's carbon footprint: the Israeli experience. Water (Switzerland) 10(2). https://doi.org/10.3390/w10020197

Trappe JM, Kovács GM, Claridge AW (2010) Validation of the new combination Mattirolomyces austroafricanus. Mycol Progr 9:145. https://doi.org/10.1007/s11557-009-0631-3

Villafañe VE, Valiñas MS, Cabrerizo MJ, Helbling EW (2015) Physio-ecological responses of Patagonian coastal marine phytoplankton in a scenario of global change: role of acidification, nutrients and solar UVR. J Mar Chem 177:-411

Wahab HA, Manap MZIA, Ismail AE, Pauline O, Ismon M, Zainulabidin MH, Mat Noor F, Mohamad Z (2019) Investigation of temperature and humidity control system for mushroom house. Int J Integrated Eng 11(6):27–37

Waldron HN, Monteiro PMS, Swart NC (2009) Carbon export and sequestration in the southern Benguela upwelling system: lower and upper estimates. Ocean Sci 5:711–718

Watanabe Y, Itanna F, Izumi Y, Awala SK, Fujioka Y, Tsuchiya K, Iijima M (2019) Cattle manure and intercropping effects on soil properties and growth and yield of pearl millet and cowpea in Namibia. J Crop Improv 33(3):395–409

Wei L, Wang Q, Wu H, Ji C, Zhao J (2014) Proteomic and Metabolomic responses of pacific oyster *Crassostrea gigas* to elevate pCO2 exposure. J Proteome 112:83–94

WHO (2009) Potassium in drinking-water. Background document for preparation of WHO. Guidelines for drinking-water quality. Geneva: World Health Organization (WHO/HSE//09.01/7)

Wilhelm M (2012) The impact of climate change in Namibia- a case study of Omusati region. Ounongo Repository:1–101. http://ir.nust.na/handle/10628/381

Zimmermann I, Matzopoulos R, Kwaambwa HM (2017) Options to improve soil fertility with national resources. Namibian J Environ 1(B):7–15

9

GIS-Based Assessment of Solar Energy Harvesting Sites and Electricity Generation Potential in Zambia

Mabvuto Mwanza and Koray Ulgen

Contents

Introduction .
   Geographical Description .
Solar Photovoltaic Power Plant Sitting Considerations
   Environmental and Social Issues
   Restricting Issues
Potential Site Identification and Mapping ...
   Solar PV Potential Site Identification and Mapping
   Available Land Area
   Electricity Generation Potential
Potential Site and Electricity Generation Potential
   Solar PV Potential Sites and Mapping
   Available Suitable Land Area ...
   Electrical Power and Electricity Generation Potential
Conclusion
References

Abstract

Land and environment are some of limited nature resource for any particular country and requires best use. Therefore, for sustainable energy generation it is often important to maximize land use and avoid or minimize environmental and social impact when selecting the potential locations for solar energy harvesting. This chapter presents an approach for identifying and determining the potential sites and available land areas for solar energy harvesting. Hence, the restricting and enhancing parameters that influence sites selection based on international

M. Mwanza (✉)
Department of Electrical and Electronic Engineering, School of Engineering, University of Zambia, Lusaka, Zambia

K. Ulgen
Ege University, Solar Energy Institute, Bornova/Izmir, Turkey
e-mail: koray.ulgen@ege.edu.tr

regulation have been imposed to the Laws of Zambia on environmental protection and pollution control legislative framework. Thus, both international regulations and local environmental protection and pollution control legislative have been used for identifying the potential sites and evaluating solar PV electricity generation potential in these potential sites. The restricting parameters were applied to reduce territory areas to feasible potential sites and available areas that are suitable for solar energy harvesting. The assessment involved two different models: firstly the assessment of potential sites and mapping using GIS, and secondly, evaluation of the available suitable land areas and feasible solar PV electricity generation potential in each provinces using analytical methods. The total available suitable area of the potential sites is estimated at 82,564.601 km^2 representing 10.97% of Zambia's total surface area. This potential is equivalent to 10,240.73 TWh annual electricity generation potential with potential to reduce CO_2 emissions in the nation and achieve SDGs. The identification of potential sites and solar energy will help improve the understanding of the potential solar energy can contribute to achieving sustainable national energy mix in Zambia. Furthermore, it will help the government in setting up tangible energy targets and effective integration of solar PV systems into national energy mix.

Keywords

Sustainable systems · Potential sites · Solar energy harvesting · Renewable energy · Zambia

Introduction

The purpose of meeting human basic needs and curbing climate change by reducing greenhouse gas emissions both at local and global levels has led to search for and establishment of energy policies for promoting renewable energy (Samuel and Owusu 2016; Sanchez-Lozano and García-Cascales 2014). The energy policies not only emphasized on promoting renewable energies but also on protecting natural resources and supporting natural environmental sustainability (Ivan 2015). Electricity generation from solar energy is in constant increase across the globe, but its share in the total energy production locally and globally still remains low as compared to fossil fuels. However, due to continual PV price decrease, increase in efficiency and maturity of technology in the last decades, feed-in tariffs including other incentives in many countries, has led to remarkable boom in photovoltaic (PV) technologies deployment and development both at utility-scale and residential levels across the globe (Robert 2014). According to International Energy Agency (IEA), the production of electricity from solar energy is expected to continue growing up to between 20% and 25% by 2050 (SEFI/UNEP 2009; Yassine 2011; Yassine and Adel 2012).

Despite of the remarkable boom in the application of solar PV technologies across the world, the application of these technologies in the electricity production in many developing countries like Zambia is still very negligible (Bowa 2017). However, there are only a few examples of small isolated solar systems used by communities, schools, companies, private households, hospitals, and health centers. These systems are often used to meet the daily energy needs and to cover up energy needs during load-shedding period (MMWED 2008; Bowa 2017). One of the largest solar systems installed by government so far through Rural Electrification Authorities (REA) was built in 2010 in Samfya district Northern Province (installed capacity of 60 kW) (Bowa 2017). According to Bowa (2017), the estimated total installed capacity of solar photovoltaic-based power plants as of 2016 was more than 2 MW (small off-grid systems). Hence, despite of the country being located in most favorable solar belt (MMWED 2008) and receiving significant higher solar irradiation than most of world's largest solar energy utilizing countries, solar energy application for electricity generation has remained negligible. According to Meteorological Department of Zambia, the country has monthly average solar radiation incident rate of 5.5 kWh/m^2-day (Gauri 2013; MMWED 2008; Walimwipi 2012; IRENA 2013). The solar radiation intensity across the country varies with western part of country having the highest annual average of approximately 5.86 kWh/m^2-day and the eastern part with the lowest of 5.68 kWh/m^2-day as shown in Fig. 1. Therefore, Zambia has a favorable climate conditions for utilization of solar energy for both production of electricity and thermal use. The total annual average global solar radiation ranges from 1981 kWh/m^2 in parts of North-Western, Eastern, Northern, Central, and Southern provinces to 2281 kWh/m^2 in parts of Luapula, Northern, and Western provinces of Zambia as illustrated in Fig. 2.

In order to increase access to electricity for all, the Government of Republic of Zambia has set targets and plans to encourage deployment and development of renewable energy facilities across the country, with hydropower and solar energy based on photovoltaic technologies expected to experience the greatest growth. However, despite of several tools being available across the globe for estimating the solar energy potential for particular location, these tools do not fully take into consideration the environmental and social issues. In addition, the surface land areas and the natural environment are some of the world scarce natural resources that require selection of the best use of these rare resources (Ronald 2016). Therefore, in order to safeguard the natural environment and consider best use of available surface land areas, energy planning and site selection for promotion and deployment of renewable energy technologies in individual countries has become one of the most challenging aspect more especially in developing countries like Zambia.

In addition, unified planning and poor site selection for intermitted renewable energy source based power plant have resulted in mismatch between the grid capacity and PV power plant output during peak time in some parts of the world (Siheng et al. 2016; Ming 2015). On the other hand, arbitrary site selection and neglecting the transmission line available reserve margin in the procedure have resulted in some PV power plant exceeding the local transmission line reserve margin and grid unable to transmit the energy to the load centers during peak hours (Chinairn 2013; Aly Sanoh 2014; Quansah 2016). Therefore, preliminary estimation and mapping of potential sites, available areas, and technical energy yield potential for intermitted renewable energy source based power plant

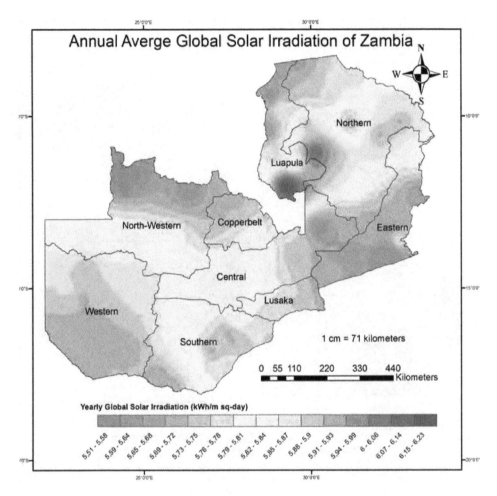

Fig. 1 Annual average horizontal solar irradiation

deployment while considering social acceptability and supporting natural environmental sustainability can be helpful to overcome these problems (Siheng et al. 2016). Doing so also helps to avoid and minimize potential negative environmental and social impacts associated with deployment of these technologies. The preliminary estimates and mapping of potential sites and technical energy yield potential for solar photovoltaic power plant development, however, have not been made in most developing countries like Zambia due to various reasons.

However, selection of potential site and evaluation of technical electricity generation potential requires a number of finer spatial resolution data, since not all locations of any particular country are suitable for deployment of these technologies due to local landscape terrain, climate, and environmental regulations (Suri 2005).

This chapter aims at providing a method for identifying and mapping a series of the potential sites and the available land areas suitable for solar energy harvesting in Zambia. The chapter further provides a method for assessing the electricity generation potential from solar energy based on commercially available solar photovoltaic technologies and available land areas. The evaluations in this chapter considered the

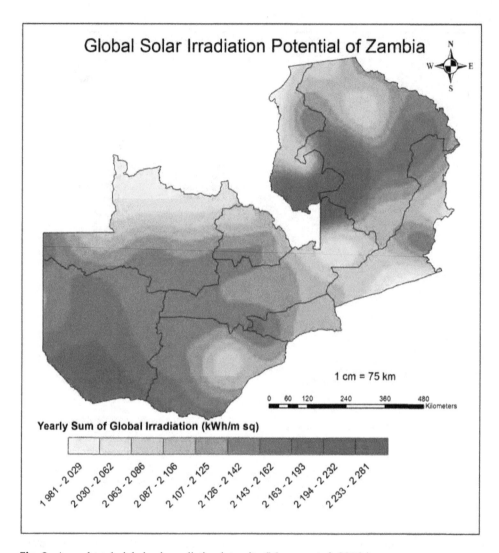

Fig. 2 Annual total global solar radiation intensity (Mwanza et al. 2016a)

modules of the solar PV systems mounted at optimal tilt position to the ground. The analysis focused on solar radiation, available areas, and typical energy that can be generated from the PV system considering the solar PV module characteristics and available solar radiation of the potential sites. The results of this study are important as it provides summarized information with regard to suitable potential sites, available land area, and technical electricity generation potential that can be attained from using solar photovoltaic technologies across Zambia.

Geographical Description

Zambia is unique country endowed with variety and abundant nature resources, such as wildlife resources, watercourse resources, forests resources, minerals resources, and renewable energy resources. Its abundant renewable energy resources such as solar energy are heavily untapped. The country is also blessed with unique climate and geography of flatland in most part of the country. It is situated between latitudes 8^0 and 18^0 south of the equator and longitudes 22^0 and 34^0 east of prime meridian. The country is landlocked by eight countries, Zimbabwe and Botswana to the South, Angola to the West, Democratic Republic of Congo and Tanzania to the North, Malawi and Mozambique to the East, and Namibia to the Southwest (Mwanza et al. 2016a).

Solar Photovoltaic Power Plant Sitting Considerations

Environmental and Social Issues

Solar energy is clean, free, and unlimited renewable energy sources that can be used for variety of purposes including pumping water for irrigations, drying and preparing food, and most importantly for electricity generation. However, just like any other alternative energy supply option, solar photovoltaic technology deployments at utility-scale are not free from imposing negative effects on both the environment and society (www.energy.gov) (Wang and Prinn 2010; Union of Concerned Scientists 2015). Most of these effects depend on development size, site, and the type of technology deployed and also site selection and environmental guidance procedure. The environmental and social impacts associated with renewable energy technology development are mainly grouped as listed in Table 1 (Ahmed Aly 2017; Turlough 2017; Shifeng and Sicong 2015; Kaoshan et al. 2015; Fylladitakis 2015; Saidur et al. 2011; Gipe 1995; Interior Department 2010; Damon and Vasilis 2011; U.S 2016; Geoffrey and Tidwell 2013; England 2011; Tsoutsos 2005, 2009).

Restricting Issues

The potential impacts associated with utilization of renewable energy technology have potential to hinder or delay deployment and development of solar photovoltaic technologies or facilities in potential sites. Table 2 and Figs.3, 4, 5, 6, 7, 8, 9, 10, 11, and 12 list solar PV systems deployment restricting issues that have, among others, been considered for inclusion, as appropriate, in the available land area, technical electricity generation potential, and potential sites assessment for sustainable solar photovoltaic facilities development in Zambia based on highlighted environmental and social impacts (www.energy.gov; Abdolvahhab Fetanat 2015; Alami et al. 2014; Ahmed et al. 2017; Arthur Bossavy 2016; Addisu and Mekonnen 2015; Marcos Rodriques 2010; Anthony Lopez 2012). Restricting criteria data are features that

Table 1 Summarized environmental and social impacts induced by solar photovoltaic power Plants

Type of impact	Causes	Factors contributing	Effects
Noise pollution	Inverter noise due to internal electronics, transformer noise, construction noise	Air temperature, humidity, ground surface material, background noise, heavy machinery, human activity	Sleep disturbance, hearing losses, headaches, irritability, fatigue, constrict arteries, weaken immune systems, annoyance, or dissatisfaction
Air and water pollution	PV module: Toxic materials, heavy machinery & transformers: Oil, switchgear breaker gas SF_6, ground clearing and grading	Fire, module cracking, leaking machinery due to poor maintenance, leakages in switchgear, transformers, & machinery, access roads & ground preparing	Death, injuries, loss of ecosystem, contamination of soil, water, and air
Water use	Periodic maintenance: PV module surface cleaning, construction activity, dust abatement activities	Dust, wind, location, size of facility, system performance, unpaved roads	Reduced underground water recharge, reduced surface water flow, agriculture water problem, wildlife water problem, human water problem
Climate Change & Greenhouse gas (GHG) emissions	Concrete and steel for PV array mounting foundations	Size of PV array, location of facility	CO_2 emission; global warming
Wildlife & Habitat loss	Excavation, grading, ground clearing, road & electrical grid construction	Location, landscape topography, size of facility, distance to road and electrical grid	Loss of feeding, nesting or roosting grounds for animals, birds, ecosystem disturbance, wildlife reduction
Visual impact	Distance to residential areas, size of facility, night lights at power plant, human perception	Scenic backgrounds, local landscape topography, local landscape between solar plant and viewers, location of solar farm, color of PV panels, layout of solar farm; irregular or regular, clear skies	Aesthetic effects, public health, negative perception of solar energy technologies, visual effect

(continued)

Table 1 (continued)

Type of impact	Causes	Factors contributing	Effects
Land use/ Soil & Land Degradation, fugitives dust	Power grid & access road construction, PV array foundation excavation, removal of surface plants, wastewater and oil from construction machinery, excess wastes from construction: Plastics, metal, glues, & inks	Layout of solar PV farms, location of solar PV farm, landscape topography, type of PV technology, tilt angle of modules, distance to access road and electrical grid, office wastes	Deforestation, soil erosion, loss of habitat, landslide, floods, air and water pollution, ecosystem disturbance, fugitive dust

Table 2 Restricting issue datasets

Type of impact eliminated, minimized or avoided	Site descriptions	Detail nature of sites descriptions
Water use, Wildlife & Habitat Loss	Wildlife sites	National parks and game reserves
Visual impact, Noise Pollution & Land use/degradation, fugitives dust	Community interest sites	Airfields, historical sites, archaeological sites, traditional and cultural heritage sites, national monuments sites and tourism sites, religious significance sites
Water use, land use/degradation	Agriculture sites	Crop areas and potential agriculture areas
Fugitive dust, water use, air/Noise Pollution & Visual Impact, land use	Settlement sites	Rural/urban and residential areas: Towns, cities, villages, and areas used extensively for recreation and aesthetic reasons
Water pollution, Wildlife & Habitat Loss, land degradation	Surface water bodies and surrounding sites	Rivers, lakes, streams, waterfalls, and wetlands
Land degradation, Wildlife & Habitat Loss	Landscape	Land elevation and slope (>5degrees), areas prone to flooding and natural hazards, zones prone to erosion or desertification, zones of high biological diversity, areas supporting populations of rare and endangered species,
Land use, Wildlife & Habitat Loss, fugitive dust, water, air and soil pollution, visual impact	Right of way	Transmission, roads and railroads network right of way
Wildlife & Habitat Loss, climate Change & Greenhouse gas (GHG) emissions, land degradation	Forest and surrounding sites	Forests: Low need-leaved deciduous forest and moderate evergreen deciduous forest, shrub-lands: Closed to open shrub-land and open shrub-land, grassland: Sparse grassland, indigenous forest

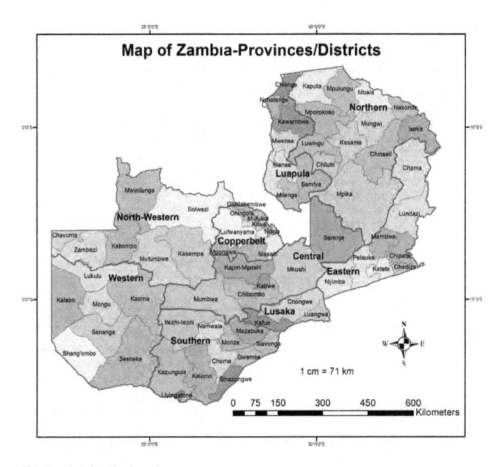

Fig. 3 Administrative boundary map

pose restrictions or limitations, that is, unsuitable or not preferred areas based on legislative laws of the country and nature.

Potential Site Identification and Mapping

Solar PV Potential Site Identification and Mapping

In order to assess the potential sites suitable for utility-scale solar photovoltaic deployment based on literatures surveyed and the laws of Zambia on environmental for development of any industry or plant on a particular site and restrictions datasets summarized in Table 2. Thus, the following environmental and social impacts and issues illustrated in Table 3, among others, are considered for inclusion, as appropriate, in the selection of suitable sites for solar energy facilities (ECZ 1994).

These maps included land elevation map (DEM), land use/cover layer map, town and village location map, community interest sites map, national parks map, surface water bodies map, roads and railway map, study area boundaries, and transmission

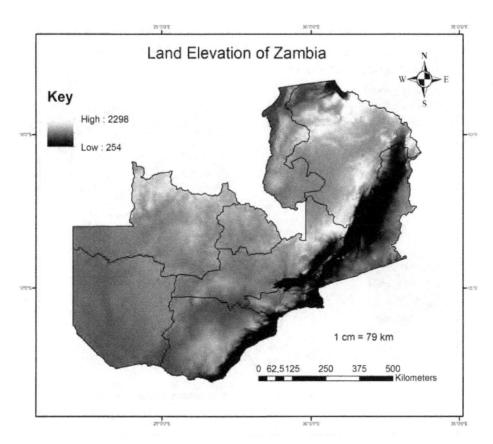

Fig. 4 Land digital elevation map

line maps (Nazli Yonca 2010; Sanchez Lozano et al. 2013, Brewer 2014, Chao-Rong Chen 2014, Charabi and Gastli 2011, Lopez 2012, Janke 2010; Uyan 2013). The rationale used for each restrictions are as follows:

- *Land Use/Cover (C6, C9):* This dataset has 10 classes including bare land, closed to open shrubland, open shrubland, sparse grassland, croplands, urban settlement, water courses, wetland, and forest sites; low need-leaved deciduous forest and moderate evergreen forest. For deployment of solar PV power plants only bare lands, sparse grassland and open shrubland were considered suitable due to easy accessibility considering an emerging economy and also to reduce land clearing costs.
- *Wildlife Sites (C2):* this dataset considers areas such as national parks, game reserves, and other natural resources since development in these sites will have adverse impact on birds, animals, and ecology, thus any construction in these areas may face public and international resistance. Therefore, these areas and the surrounding areas within the buffer of 2 km were considered not suitable (Nazli Yonca 2010).

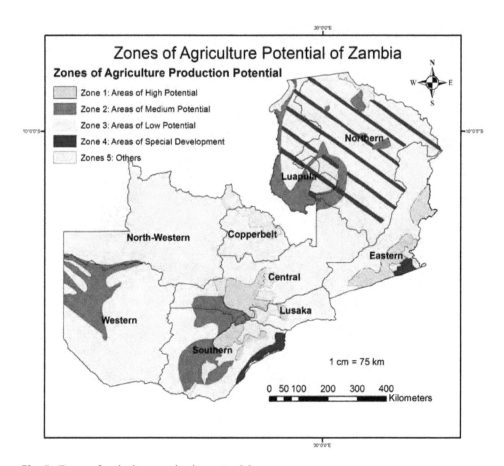

Fig. 5 Zones of agriculture production potential

- **Settlement and Community Interest Sites (C4, C1):** The dataset consists of settlement areas for both rural and urban such as airfields, airports, towns, villages, and other dwelling areas and community interest sites. Here a buffer of 3 km is considered to avoid aforementioned impacts and increase public safety and acceptance. All areas outside the buffer were considered suitable (Joss and Watson 2015).
- **Land Elevation (C7):** As it is expected that no one will install solar PV power plants in gorges or higher elevation due to construction costs. Thus, this dataset considered all higher and lower elevations such as mountains and gorges with steeper slopes above 5^0 as unsuitable areas.
- **Surface Water Bodies (C8):** In this dataset all surface water bodies such as rivers, streams, lakes, including waterfalls, and wetlands were considered as protected areas in order to avoid water pollution. Thus, a buffer of 2 km was considered with all areas outside buffer being suitable.
- **Roads and Railways Network (C3):** The dataset considers roads and railway network to be restriction since no one is supposed to build on roads or railway and

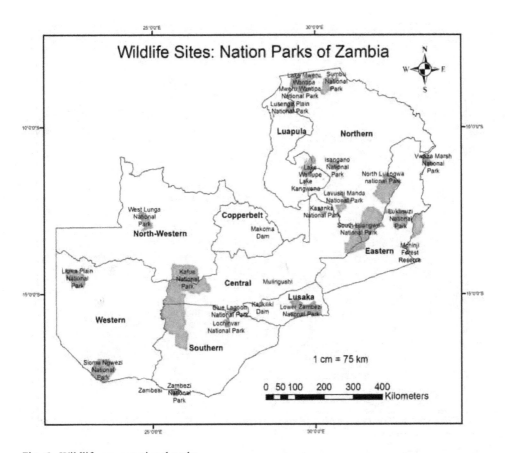

Fig. 6 Wildlife areas: national parks

also for the safety of the public. Hence, a 0.5 km buffer has been considered in order to increase public safety and also reduce cost of constructing access road which usually leads to land use/degradation, wildlife and habitat loss, fugitive dust, and air and soil pollution to the site and surrounding areas. Thus, the areas outside the buffer are considered suitable.

• **Transmission Line Network (C5):** In this dataset the right of way for transmission line were considered as unsuitable area for solar PV power plants, thus a 0.5 km buffer was used. Any areas within the buffer were considered unsuitable. The 0.5 km buffer were considered so that the cost of constructing new transmission lines is reduced, but at the same time to avoid conflict with right of way for transmission lines and avoid land use/degradation, wildlife and habitat loss, fugitive dust, water, and air and soil pollution to the site and surrounding areas.

After creating buffers, and changing some features from vector to raster, in order to evaluate available areas and identify/map feasible potential sites, the created buffers for the restricting layers were overlaid on each other using GIS spatial analysis. Figure 13 below shows the summarized analysis procedure.

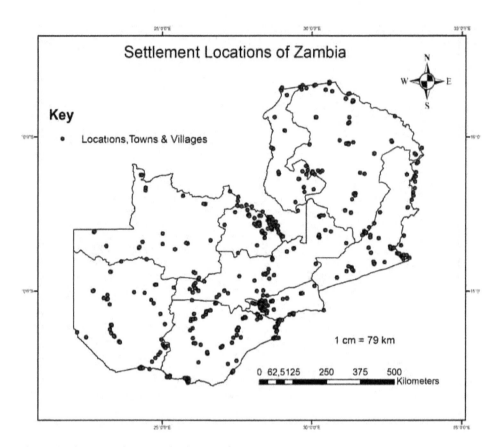

Fig. 7 Settlement and community interest sites

Available Land Area

In order to estimate the available suitable areas for solar photovoltaic power plant development based on aforementioned restrictions issues, a new factor called Area Suitability factor f_{SF} was introduced. It is defined as the ratio of total grid cells of suitable surface area to the total cells of the study surface area. The factor is estimated based on study area grid cells; here the total grid cells for study surface area are evaluated considering the sum of restricted and suitable surface areas' cells. Hence the factor depends on the ratio of available suitable area and surface area of the study area and it is calculated using the expression below.

$$f_{SF} = \left(\frac{C_{CSA}}{C_{CSA} + C_{CRA}} \right) = \frac{C_{CSA}}{C_{TSSA}} \tag{1}$$

where C_{CSA} is the total number of cells of suitable areas, C_{CRA} is the total number of cells of restricted areas, and C_{TSSA} is total number of cells of study area.

Therefore, the total available suitable land areas for each district and for Zambia were evaluated using expression 2 given below

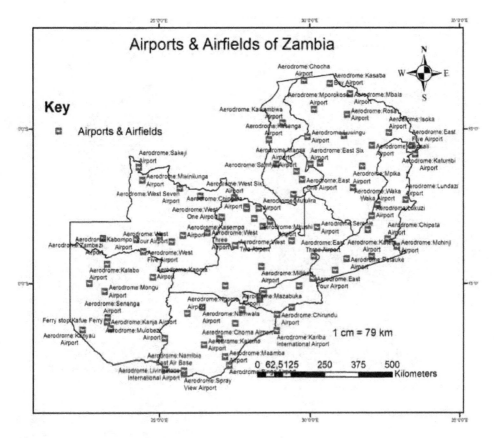

Fig. 8 Airports and airfield

$$A_{ADS} = f_{SF} \cdot A_{TSA} \tag{2}$$

where A_{ADS} is total available suitable areas (km^2), and A_{TSA} is total surface area of the study area (km^2).

Electricity Generation Potential

Solar Energy Potential in Zambia

According to the literature and data undertaken by Meteorological Department of Zambia, the country has a significant potential of solar energy for both electrical power production and thermal from solar energy technologies. The country has average peak sunshine of about 6–8 hours per day and monthly average solar radiation of 5.5 kWh/m^2-day throughout the year (MMWED 2008; Walimwipi 2012). According to International Renewable Energy Agency (IRENA 2012), the country has the highest total yearly solar radiation of 2,750 kWh/m^2-year with the highest average temperature of 30 °C, which presents good opportunity for solar systems deployment (IRENA 2012).

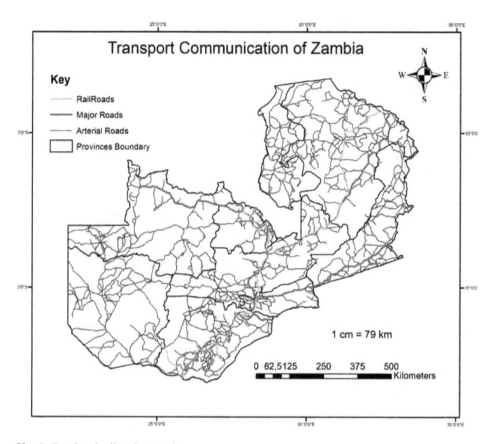

Fig. 9 Road and railroad networks

Performance of PV System

In order to evaluate the performance of grid connected PV power plants, the following performance indices are normally used: yields, normalized losses, and system efficiencies, performance ratio, and capacity factor – (British Standard 1998). However, in this chapter final yield, performance ratio and capacity factor have been adopted for analyzing the PV system performance of the various types of PV technologies commercially available on the market (Table 3) considering Zambia's weather condition. In addition, several PV technologies have been considered in the evaluation of technical electricity generation and power potential: firstly, because the energy generation by PV power plants with same peak power and receiving same amount of solar irradiation differs depending on the type of technology employed in the power plants, and secondly, the amount of peak power that can be installed at a given land area differ with PV technologies as shown in Table 4. Hence, it can be concluded that the type of cell technology has greater influence in the amount of land area needed for a peak power installation, the higher the efficiency the lower the land requirements for the peak power capacity installation (Martin-Chivelet 2016).

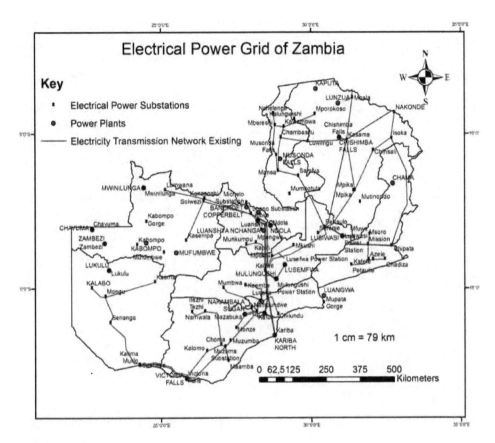

Fig. 10 Existing electrical network

Energy Model of PV Array

The solar energy resources are the key determinants of PV system electricity generation (IRENA 2012). The higher the solar energy resources, the more output yield of a PV systems per kilowatt. However, higher temperatures, dust, shading, balance of system inefficiencies, and PV technology characteristics have negative impact on the PV system energy yield (Didler 2012). Therefore, the electricity generated and supplied to grid by PV system considering these negative impacts has been estimated using Eq. 3:

$$E_A = A_{PV} \cdot H_R \cdot \eta_P (1 - \lambda_p)(1 - \lambda_C) \tag{3}$$

where E_A is energy output of PV system (MWh/year), H_R is solar radiation on the surface of module (kWh/m^2-day), A_{PV} is PV system active area (m^2), η_P is module efficiency under STC condition, λ_p is miscellaneous module losses due to dusty covering, and λ_C is losses due to power conditioning unit and cable losses. Module efficiency is a function of its nominal efficiency, η_r, which is measured at STC $T_r = 25\ °C$ (Didler 2012). İt has been calculated as:

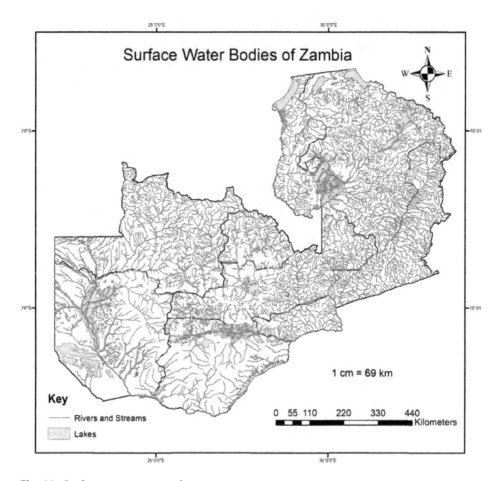

Fig. 11 Surface watercourses and streams

$$\eta_p = \eta_r \cdot [1 - \beta(T_c - T_r)] \tag{4}$$

where β is a temperature coefficient for module efficiency, T_c is a module temperature due to air temperature, and T_r is STC reference temperature.

Module temperature is related to the average monthly ambient air temperature, Ta, for a local condition has been calculated using Eq. 5 (Didler 2012).

$$T_c; = T_a + \frac{G_T}{G_{T.NOCT}} \left(\frac{9.5}{5.7 + 3.8V_W} \right) (T_{c,NOCT} - 20)(1 - \eta_m) \tag{5}$$

where G_T is solar irradiance (W/m^2), T_a is ambient air temperature (°C), and V_W is wind speed(m/s) for the location, $T_{c,NOCT}$ is nominal operating cell temperature (Table 3), it depends on type of PV technology, η_m is the factor less than 1 and normally neglected and $G_{T,NOCT}$ is 800 W/m^2.

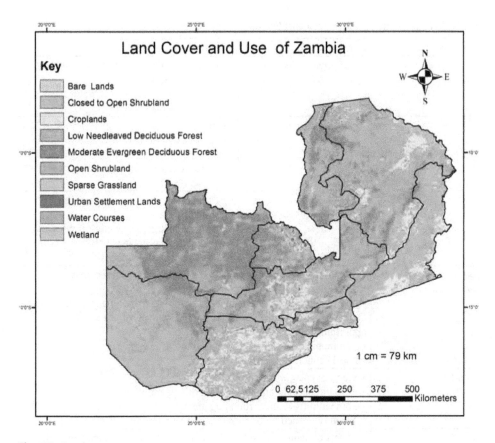

Fig. 12 Land aspects

Performance Ratio Model

Performance ratio is denoted by PR, this factor is important as it shows the overall effect of losses on the PV array's rated output power due to the PV array temperature, incomplete use of the solar irradiation, and PV system component inefficiencies or failures. It is calculated as (British Standard 1998).

$$PR_{.} = \frac{Y_F}{Y_R} = \frac{E_{AC}}{G\eta_{STC}} \tag{6}$$

where G-standard test condition solar radiance (1 kW/m^2) and η_{STC}-array efficiency at standard test condition given as.

$$\eta_{STC} = \frac{P_{PV}}{GA_{PV.A}} \tag{7}$$

where $A_{PV,A}$-Active array area (m^2) and $P_{PV,A}$- array rated power (kW$_P$).

Table 3 Environmental and social consideration for site selection in Zambia (ECZ 1994)

Issues	Considerations	Effect description
Ecological	(a) Biological diversity.	• Effect on number, diversity, breeding sites, etc. of flora and fauna • Effect on the gene pools of domesticated and wild sustainable yield
	(b) Sustainable use including:	• Effect of soil fertility • Breeding populations of fish and wildlife (game) • Natural regeneration of woodland and sustainable yield
	(c) Ecosystem maintenance including:	• Effects of proposal on food chains • Nutrient cycles • Aquifer recharge, water run-off rates, etc. • Aerial extent of habitats • Biogeographical processes
Social, economic, and cultural	• Effects of generation or reduction of employment in the area • Social cohesion or disruption (resettlement) • Immigration (including induced development when people are attracted to a development site because of possible enhanced economic opportunities) • Communication-roads opened up, closed, re-routed • Local economic impacts	
Land scope	• Views opened up or closed • Visual impacts (features, removal of vegetation, etc.) • Compatibility with surrounding areas • Amenity opened up or closed, e.g., recreation facilities	
Land use	• Effects on land uses and land potential in the project area and in the surrounding areas • Possibility of multiple use	
Water	• Effects on surface water quality and quantity • Effects on underground water quality and quantity • Effect on the flow regime the water course	
Air quality	• Effects on the quality of the ambient air of the area • Type and amount of possible emissions (pollutants)	

Capacity Factor Model

This is a model used to show the amount of energy delivered to the grid by an electric power generation system (Ayompe 2014). It is defined as the ratio of the output actual annual energy generated by PV system to the amount of energy the PV system would generate if it is operated continuously at full rated power for 8,760 hours in a year and it is expressed as (Ayompe 2014; Kynakis 2009; British Standard 1998).

$$CF. = \frac{E_{AC}}{8760 \times P_{PV}} = \frac{PR \times H_t}{8760 \times P_{PV}} \quad (8)$$

where CF is capacity factor (%), E_{AC} is Actual annual energy output (kWh/year), and P_{PV} is Full rated PV power (kW$_p$).

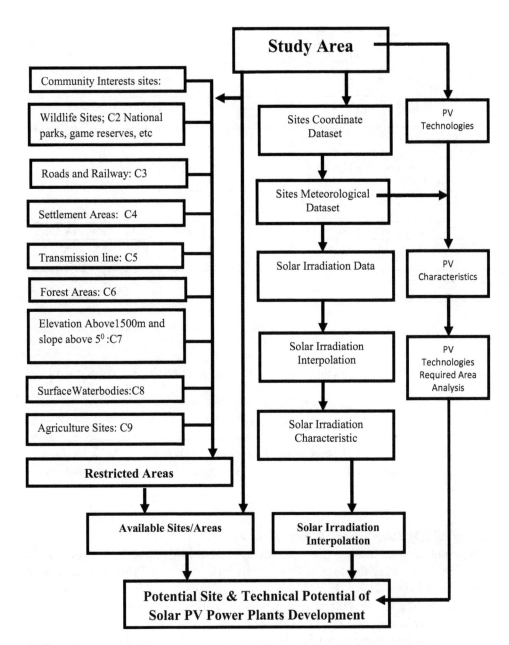

Fig. 13 Analysis Procedure

Solar Energy Potential Model

Theoretical Solar Energy Potential Model

Theoretical solar energy potential involves the assessment of the total solar energy that is received at the surface of the study area. This potential involves identifying the study area boundary and the size of the study land area, including total annual

Table 4 PV technology parameters

PV technology parameters	PV technologies					Reference
	mc-Si	Pc-Si	a-Si	CIS	CdTe	
Efficiency (%)	23	16	7–10	12.1	11.2	IRENA 2012
Temp. Coeff. β (%/^{0}C)	0.41	0.43	0.27	0.26	0.25	Suprava and Pradip 2015
Active PV area needed (m^2/kW)	7	8	15	10	11	IRENA 2012
Total PV system area needed (m^2/MW)	14,000	16,000	30,000	20,000	220,000	Estimated
NOCT (°C)	47	45	40.3	45.6	45	Suprava and Pradip 2015
Max. PV module (W)	320	320	300	120	120	IRENA 2012
BOS losses (%)	8	8	8.13	11.33	11.33	Various sources
Dust factor (%)	5	5	5	5	5	Various sources

average solar radiation magnitude. The theoretical potential has been calculated using Eq. 9:

$$E_{TH} = A_S \cdot H_P \qquad (9)$$

where E_{TH} is theoretical solar energy potential (MWh/year), A_s is study area active surface area (km^2), and H_R is total annual average solar irradiance (MWh/km^2-year).

Geographical Solar Energy Potential Model

Geographical solar energy potential involves assessing the solar energy that is received on the available and suitable land area of the active surface land area of study area (Lopez 2012). Hence, the process of assessing this potential involved firstly excluding all the protected and restricted areas from the active surface area of the study area under consideration (Yan-wei 2013; Lopez 2012).

Therefore, the remaining surface land area is taken as the most suitable land area of the total study area surface land area for solar energy technologies development. In this study, the geographical solar energy potential has been estimated using Eq. 10 given below:

$$E_G = A_{AOS} . H_R. \qquad (10)$$

where E_G is geographical solar energy potential (kWh/year), A_{ADS} is Available Suitable Area (m^2), and H_R is total annual average solar radiation (kWh/m^2-year).

Solar PV Technical Power Potential Model

The process of assessing the feasible solar PV technical potential, that is, the maximum power capacity that can be installed for any country without environmental and social impacts involves firstly by excluding restricted areas and areas not suitable for utility-scale PV systems development within the defined boundaries. Furthermore, considering technical characteristics of solar PV technologies (Table 3) to convert the solar energy to electrical energy, the total solar energy that is received at the surface of the solar PV module and the area required by the PV system and its supporting infrastructures. Hence, the technical solar PV potential has been estimated using Eq. 11 (Yan-wei 2013; Lopez 2012):

$$P_{TP} = \left(\frac{A_{ADS}}{A_{PVSA}}\right). = P_{PD}A_{PV} \qquad (11)$$

where P_{TP} is Solar PV Power Potential (MW), A_{PVSA} is Solar PV system and Supporting Infrastructure Occupied Area per MW (km^2/MW), A_{ADS} is Available Suitable Area for Study Area (km^2), A_{PV} is total geographical occupied area by PV system and supporting infrastructure (km^2), and P_{PD} is solar power density of the area (MW/km^2).

Solar PV Systems Electricity Generation Technical Potential Model

The total AC electricity generated by the PV system is the sum of the electricity produced by all array in the PV power plant measured at the point where the system fed to utility grid. The total daily $E_{AC,DP}$ and monthly $E_{AC,mP}$ AC energy generated by plant are expressed as (Ali et al. 2016; Tripathi et al. 2014; Siyasankari and Babu 2015):

$$E_{AC,DP} = \sum_{t-1}^{24} E_{AC,t} \qquad (12)$$

$$E_{AC,mP} = \sum_{d-1}^{N} E_{AC,DP} \qquad (13)$$

where N is number of days in the month, and $E_{AC,t}$ is energy produced by PV power plant per hour (kWh).

Utility-scale photovoltaic are large-scale solar PV power plant that can be deployed within the boundaries of the country on an open space land area (Lopez 2012). Several studies have considered that the modules covers the available suitable areas on horizontal; however, the method proposed in this study seeks to consider the active area of PV arrays only and also the supporting infrastructures in the evaluation of technical potential. The process of assessing the extractable electricity generation potential from the sun for any country involves firstly by excluding areas not suitable for utility-scale PV systems within the defined boundaries, and secondly, considering technical characteristics of PV systems to convert the solar energy to electrical energy and the area required by the PV system and its supporting infrastructures. In this study the technical solar energy potential was estimated using Eq. 14 (Yan-wei 2013; Lopez 2012):

$$E_T = P_{TP}.CF.T_{TSH} \qquad (14)$$

where E_T is Solar PV Electricity Generation Potential (MWh/year), P_{TP} is the technical power potential (MW), CF is Study Area Capacity factor (%), and T_{TSH} is the hours of the whole year (8,760 hours/year).

Potential Site and Electricity Generation Potential

Solar PV Potential Sites and Mapping

Figure 14 presents the map of solar PV potential suitable sites evaluated for Zambia, which indicates that the country has large land areas suitable for solar PV power plant development both at district and provincial levels. The aim of this case was focused on mapping the potential sites suitable for PV power plant installation with minimized or no environmental and social impacts. Therefore, all limiting factors considered not suitable for PV systems and those areas likely to have environmental and social issues were eliminated in the analysis using GIS spatial analysis. Hence, the Solar PV Potential Sites atlas shows that the country has the largest suitable site for solar PV power plant development in the Southern Province with Lusaka Province having the least. However, it can be observed that the available suitable areas are distributed throughout the country, hence providing opportunity for wide deployment of the solar PV technologies across the country. In addition, the atlas also shows that regions near to the national power grid contain suitable sites for easy integration of these technologies into the national energy mix and national power grid. The atlas provides essential information for sites close to villages and towns far from the grid offering opportunity for mini off-grid systems. Therefore, the atlas offers vital information for setting targets for electrification of both rural and urban areas of the country.

Available Suitable Land Area

Table 5 shows the annual average solar irradiation, total surface area and the available suitable areas for each district of Zambia. This reveals significant differences in suitable available areas within the 75 districts and 9 provinces across the country due to the availability of the aforementioned restricting factors considered in the evaluation. It can be observed that the districts in Eastern Province have lowest ratios of suitable area to surface areas in the ranges of 1.57 to 11.61% mainly due to the availability of restricting factors such as escapement, protected areas (e.g., National Parks, Zones of higher Agriculture Potential), and agriculture activities.

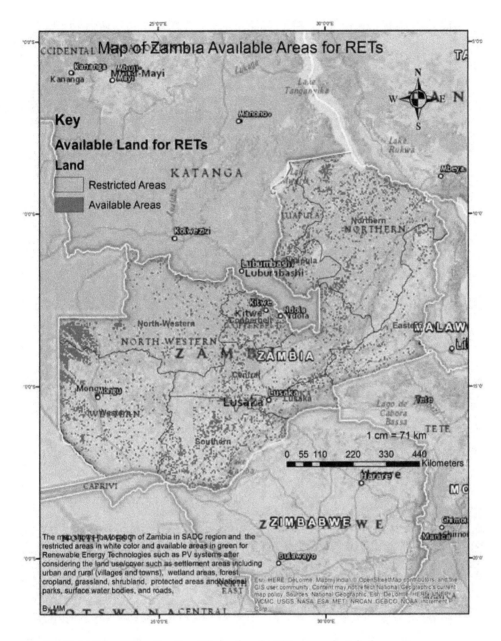

Fig. 14 Potential suitable sites for utility-scale solar PV power plant deployment

The provincial total suitable areas available for utility-scale solar photovoltaic power plants development as shown in Table 6 ranged from 2,151.70 km^2 (Lusaka) to 16,593.56 km^2 (Southern). As earlier stated, Eastern Province has the lowest annual average solar irradiation and also the overall percent suitable area (6.61%) whereas Southern Province has the largest (19.33%). However, Lusaka Province due to its size and population has the lowest overall suitable area (2,151.70 km^2) followed by Copperbelt (4,475.66 km^2) and highest being the Southern Province

Table 5 Available suitable areas in the districts of Zambia

Province	Districts / Sites	Coordinates S (°)	E (°)	Annual solar irradiation (kWh/m²-day)	Total surface area (km²)	Percentage suitable area (%)	Total suitable area (km²)
Eastern	Chama	11.216	33.162	5.79	18,152	8.07	1,464.8664
	Chipata	13.641	32.646	5.78	6,172	1.71	105.5412
	Chadiza	14.061	32.417	5.58	2,541	4.72	119.9352
	Petauke	14.248	31.322	5.65	8,495	11.61	986.2695
	Katete	14.051	32.047	5.59	3,969	6.86	272.2734
	Lundazi	12.281	33.173	5.65	13,517	1.57	212.2169
	Nyimba	14.557	30.822	5.69	10,444	10.41	1,087.2204
	Mambwe	13.550	31.756	5.71	5,918	5.47	323.7146
Total	-	-	-	-	**69,208**	**6.61**	**4,572.0376**
Lusaka	Chongwe	15.336	28.670	5.72	11,728	11.45	1,342.856
	Kafue	15.765	28.181	5.72	5,658	10.39	587.8662
	Luangwa	15.305	30.037	5.64	4,062	5.00	203.1000
	Lusaka	15.403	28.287	5.72	447	4.00	17.8800
Total	-	-	-	-	**21,896**	**9.83**	**2,151.7022**
Southern	Choma	16.805	27.004	5.71	7,010	18.54	1,299.654
	Gwembwe	16.494	27.598	5.71	4,048	11.81	478.0688
	Itezhi-tezhi	15.734	26.054	5.83	16,310	23.78	3,878.518
	Kalomo	17.025	26.477	5.80	13,808	22.71	3,135.7968
	Kazungula	17.550	25.425	5.92	18,375	23.78	4,369.575
	Livingstone	17.843	25.840	5.92	755	12.16	91.8080
	Mazabuka	15.856	27.751	5.83	6,432	17.30	1,112.736
	Monze	16.278	27.488	5.71	4,685	14.47	677.9195
	Namwala	15.739	26.455	5.83	5,216	5.16	269.1456
	Siavonga	16.502	28.746	5.75	4,284	17.31	741.5604
	Sinazongwe	17.220	27.477	5.77	4,898	11.00	538.7800

(continued)

Table 5 (continued)

Province	Districts	Coordinates		Annual solar irradiation	Total surface area	Percentage suitable area	Total suitable area
Total	-	-	-	-	**85,823**	**19.33**	**16,593.5621**
Luapula	Chienge	8.653	29.166	5.67	3,391	5.77	195.6607
	Kawambwa	9.800	29.078	5.78	9,651	8.67	836.7417
	Mansa	11.197	28.893	5.76	10,096	23.77	2,399.8192
	Milenge	12.082	29.295	6.23	5,930	9.64	571.6520
	Mwense	10.392	28.666	5.63	6,654	11.80	785.1720
	Nchelenge	9.361	28.742	5.62	3,632	7.46	270.9472
	Samfya	11.353	29.491	5.80	11,213	8.98	1,006.9274
Total	-	-	-	-	**50,567**	**11.998**	**6,066.9202**
North Western	Chavuma	13.079	22.681	5.81	4,434	47.17	2,091.5178
	Kabompo	13.593	24.203	5.79	14,295	6.15	879.1425
	Kasempa	13.459	25.840	5.80	22,061	4.28	944.2108
	Mufumbwe	13.678	24.796	5.84	19,734	4.12	813.0408
	Mwinilunga	11.738	24.428	5.51	21,191	4.26	902.7366
	Solwezi	12.168	26.384	5.60	30,232	6.44	1,946.9408
	Zambezi	13.539	23.115	5.80	13,879	28.37	3,937.4723
Total	-	-	-	-	**125,826**	**9.15**	**11,515.0616**

Available suitable areas in the districts of Zambia

Province	Districts	Coordinates		Solar irradiation	Total surface area	Percentage suitable area	Total suitable area
	-	$S(^0)$	$E(^0)$	$(kWh/m^2\text{-day})$	(km^2)	(%)	(km^2)
Northern	Chinsali	10.542	32.080	5.90	14,939	4.12	615.4868
	Chilubi	11.073	30.130	6.04	5,187	3.58	185.6946
	Isoka	10.150	32.660	5.90	9,344	5.05	471.8720
	Kaputa	8.472	29.662	5.67	12,843	16.16	2,075.4288
	Kasama	10.201	31.193	5.88	10,590	10.86	1,150.0740
	Luwingu	10.250	29.916	5.73	8,721	11.96	1,043.0316

	Mbala	8.847	31.371	5.77	8,662	11.37	984.8694
	Mpika	11.824	31.440	5.83	40,025	6.38	2,553.5950
	Mporokoso	9.373	30.125	5.75	12,028	12.44	1,496.2832
	Mpulungu	8.771	31.124	5.77	10,351	7.53	779.4303
	Mungwi	9.609	32.212	5.84	10,051	8.37	841.2687
	Nakonde	9.354	32.723	5.84	4,445	18.45	820.1025
Total	-			-	**147,186**	**8.84**	**13,017.1369**
Central	Chibombo	14.703	28.106	5.80	13,298	10.50	1,396.2900
	Kabwe	14.435	28.435	5.80	1,594	5.95	94.8430
	Kapiri-Mposhi	13.955	28.674	5.80	12,120	13.11	1,588.9320
	Mkushi	13.995	29.474	5.72	22,552	8.06	1,817.6912
	Mumbwa	15.006	27.059	5.83	21,755	10.59	2,303.8545
	Serenje	13.253	30.284	5.62	23,075	12.35	2,849.7625
Total	-			-	**94,394**	**10.65**	**10,051.3732**
Copperbelt	Chililabombwe	12.353	27.834	5.70	938	15.00	140.7000
	Chingola	12.538	27.837	5.70	1,766	13.95	246.3570
	Kalulushi	12.844	28.026	5.74	1,121	16.00	179.3600
	Kitwe	12.809	28.216	5.74	889	21.67	192.6463
	Luanshya	13.152	28.413	5.80	950	16.22	154.0900
	Lufwanyama	12.678	27.279	5.70	11,316	11.05	1,250.4180
	Masaiti	13.280	28.408	5.80	3,703	19.73	730.6019
	Mpongwe	13.529	28.144	5.82	8,465	14.71	1,245.2015
	Mufulira	12.557	28.240	5.74	1,145	15.91	182.1695
	Ndola	12.980	28.628	5.74	1,035	14.89	154.1115

(continued)

Table 5 (continued)

Province	Districts	Coordinates		Annual solar irradiation	Total surface area	Percentage suitable area	Total suitable area
Total	–	–	–	–	**31,328**	**14.29**	**4,475.6557**
Western	Kalabo	14.998	22.681	5.86	18,065	23.49	4,243.4685
	Kaoma	14.817	24.807	5.85	22,099	4.90	1,082.8510
	Lukulu	14.408	23.258	5.89	15,589	21.09	3,287.7201
	Mongu	15.274	23.148	5.90	10,125	4.00	405.0000
	Senanga	16.120	23.303	5.93	15,205	9.53	1,449.0365
	Sesheke	16.747	24.549	5.91	29,423	6.97	2,050.7831
	Shang'ombo	16.317	22.264	5.92	15,880	10.09	1,602.2920
Total	–	–	–	–	**126,386**	**11.17**	**14,121.1512**

Table 6 Provincial total suitable areas for utility-scale solar photovoltaic power plants

Province	Annual solar irradiation (kWh/m²-day)	Total area (km²)	Suitable area (km²)	Percent suitable area (%)
Lusaka	5.70	21,896	2,151.7022	9.82
Luapula	5.78	50,567	6,066.9202	12.00
Central	5.76	94,394	10,051.3732	10.65
Copperbelt	5.75	31,328	4,475.6557	14.29
Northern	5.83	147,186	13,017.1369	8.84
N/Western	5.74	125,826	11,515.0616	9.15
Western	5.89	126,386	14,121.1512	11.17
Southern	5.80	85,823	16,593.5621	19.33
Eastern	5.68	69,208	4,572.0376	6.61
Zambia	**5.78**	**752,614**	**82,564.6007**	**10.97**

(16,593.56 km²) (Fig. 15a, b). In short, comparing only available suitable areas where installation of PV system is suitable, Southern province has about 7.71 times more suitable area than Lusaka Province. However, there are large differences in surface area size between the two provinces, with Lusaka having 3.92 times less surface area than Southern Province. The country has approximately 10.97% equivalent to 82,564.60 km² of the total suitable surface land area for development of utility-scale solar PV power plant (Table 6).

Electrical Power and Electricity Generation Potential

Table 7 shows district solar energy theoretical and geographical energy potential. Since these potentials depend on the solar irradiation and available surface area and available geographical suitable areas. Hence areas with larger surfaces and receiving the higher solar irradiation such as Northern, Western, and North-Western have the highest overall theoretical potential whereas areas with larger suitable areas such as Southern, Western, Northern, North-Western, and Central Provinces have higher geographical solar energy potential (Table 8 and Fig. 16).

The district-based solar PV technical power potential by technology (Table 9) shows that crystalline silicon based solar PV technologies possess large potential due to less land requirements for installation, with monocrystalline-silicon technology having the largest technical power potential of 5,897.46 GW whereas amorphous-silicon having the lowest potential of 2,752.16 GW due to huge land requirements. The variation in power potential per district is highly depended on the available suitable areas in each district which is as a result of local geographical and terrain features.

The provincial solar PV technical power potential per technology (Table 10 and Fig. 17) shows that Southern Province, followed by Western have the highest potential and Lusaka Province being the lowest. Figure 18 shows the comparison

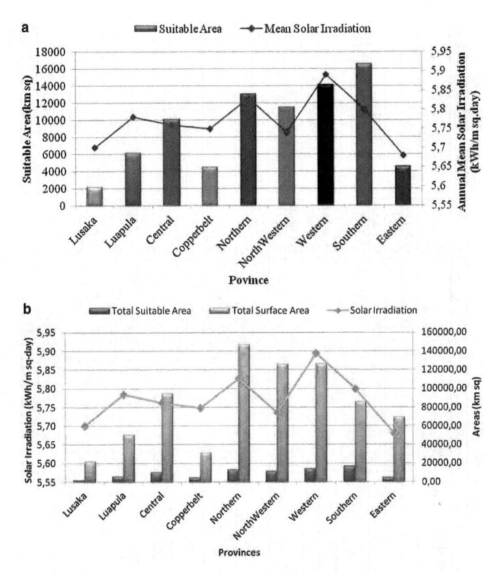

Fig. 15 (**a**) Provincial total suitable areas for utility-scale solar photovoltaic power plants. (**b**) Comparison of provincial total surface areas and suitable areas for utility-scale solar photovoltaic power plants

of solar PV technologies peak power potential for Zambia, with monocrystalline silicon having the largest whereas amorphous silicon having the lowest potential.

In absolute numbers, the highest electricity generation can be generated in the Southern, Western, Northern, North-Western, and Central Provinces due to large available suitable land areas for utility-scale solar PV system development (Table 12 and Fig. 19). Table 11 illustrates the district solar PV technical electricity generation potential by technology. Just like technical power potential it can be observed that districts with large suitable areas have the largest electricity generation potential.

Table 7 District solar energy theoretical and geographical potential

Province	Districts	Annual solar irradiation	Total surface area	Total suitable area	Theoretical potential	Geographical potential
	Sites	(kWh/m²-day)	(km²)	(km²)	(TWh/year)	(TWh/year)
Eastern	Chama	5.79	18,152	1,464.8664	38,361.53	3,095.78
	Chipata	5.78	6,172	105.5412	13,021.07	222.66
	Chadiza	5.58	2,541	119.9352	5,175.25	244.27
	Petauke	5.65	8,495	986.2695	17,518.81	2033.93
	Katete	5.59	3,969	272.2734	8,098.15	555.53
	Lundazi	5.65	13,517	212.2169	27,875.43	437.64
	Nyimba	5.69	10,444	1,087.2204	21,690.62	2,257.99
	Mambwe	5.71	5,918	323.7146	12,334.00	674.67
Total			69,208	4,572.0376	143,482.03	9,478.75
Lusaka	Chongwe	5.72	11,728	1,342.856	24,485.72	2,803.61
	Kafue	5.72	5,658	587.8662	11,812.77	1,227.35
	Luangwa	5.64	4,062	203.1000	8,362.03	418.10
	Lusaka	5.72	447	17.8800	933,25	37,33
Total			21,896	2,151.7022	45,554.63	4,476.62
Southern	Choma	5.71	7,010	1,299.654	14,609.89	2,708.67
	Gwembwe	5.71	4,048	478.0688	8436,64	996.37
	Itezhi-tezhi	5.83	16,310	3,878.518	34,706.86	8,253.29
	Kalomo	5.80	13,808	3,135.7968	29,231.54	6,638.48
	Kazungula	5.92	18,375	4,369.575	39,704.70	9,441.78
	Livingstone	5.92	755	91.8080	1,631.40	198.38
	Mazabuka	5.83	6,432	1,112.736	13,686.97	2,367.85
	Monze	5.71	4,685	677.9195	9,764.24	1,412.89
	Namwala	5.83	5,216	269.1456	11,099.39	572.73
	Siavonga	5.75	4,284	741.5604	8,991.05	1,556.35
	Sinazongwe	5.77	4,898	538.7800	10,315.43	1,134.70

(continued)

Table 7 (continued)

Province	Districts	Annual solar irradiation	Total surface area	Total suitable area	Theoretical potential	Geographical potential
Total			**85,823**	**16,593.5621**	**181,630.34**	**35,117.56**
Luapula	Chienge	5.67	3,391	195.6607	7,017.84	404.93
	Kawambwa	5.78	9,651	836.7417	20,360.71	1,765.27
	Mansa	5.76	10,096	2,399.8192	21,225.83	5,045.38
	Milenge	6.23	5,930	571.6520	13,484.52	1,299.91
	Mwense	5.63	6,654	785.1720	13,673.64	1,613.49
	Nchelenge	5.62	3,632	270.9472	7,450.32	555.79
	Samfya	5.80	11,213	1,006.9274	23,737.92	2,131.67
Total			**50,567**	**6,066.9202**	**106,760.30**	**12,808.87**
North Western	Chavuma	5.81	4,434	2,091.5178	9,402.96	4,435.38
	Kabompo	5.79	14,295	879.1425	30,210.34	1,857.94
	Kasempa	5.80	22,061	944.2108	46,703.14	1,998.89
	Mufumbwe	5.84	19,734	813.0408	42,064.99	1,733.08
	Mwinilunga	5.51	21,191	902.7366	42,618.28	1,815.54
	Solwezi	5.60	30,232	1,946.9408	61,794.21	3,979.55
	Zambezi	5.80	13,879	3,937.4723	29,381.84	8,335.63
Total			**125,826**	**11,515.0616**	**263,421.22**	**24,107.19**
Province	Districts	Solar irradiation (kWh/m^2-day)	Total surface area (km^2)	Total suitable area (km^2)	Theoretical potential (TWh/year)	Geographical potential (TWh/year)
Northern	Chinsali	5.90	14,939	615.4868	32,171.14	1,325.45
	Chilubi	6.04	5,187	185.6946	11,435.26	409.38
	Isoka	5.90	9,344	471.8720	20,122.30	1,016.18
	Kaputa	5.67	12,843	2,075.4288	26,579.23	4,295.20
	Kasama	5.88	10,590	1,150.0740	22,728.26	2,468.29
	Luwingu	5.73	8,721	1,043.0316	18,239.54	2,181.45
	Mbala	5.77	8,662	984.8694	18,242.61	2,074.18

	Mpika	5.83	40,025	2,553.5950	85,171.20	5,433.92
	Mporokoso	5.75	12,028	1,496.2832	25,243.77	3,140.32
	Mpulungu	5.77	10,351	779.4303	21,799,72	1,641.52
	Mungwi	5.84	10,051	841.2687	21,424,71	1,793.25
	Nakonde	5.84	4,445	820.1025	9,474.96	1,748.13
Total			147,186	13,017.1369	313,025.37	27,683.98
Central	Chibombo	5.80	13,298	1,396.2900	28,151.87	2,955.95
	Kabwe	5.80	1,594	94.8430	3,374.50	200.78
	Kapiri-Mposhi	5.80	12,120	1,588.9320	25,658.04	3,363.77
	Mkushi	5.72	22,552	1,817.6912	47,084.07	3,794.98
	Mumbwa	5.83	21,755	2,303.8545	46,293.55	4,902.49
	Serenje	5.62	23,075	2,849.7625	47,333.75	5,845.72
Total			94,394	10,051.3732	198,511.37	21,138.12
Copperbelt	Chililabombwe	5.70	938	140.7000	1,951.51	292.73
	Chingola	5.70	1,766	246.3570	3,674.16	512.55
	Kalulushi	5.74	1,121	179.3600	2,348.61	375.78
	Kitwe	5.74	889	192.6463	1,862.54	403.61
	Luanshya	5.80	950	154.0900	2,011.15	326.21
	Lufwanyama	5.70	11,316	1,250.4180	23,542.94	2,601.49
	Masaiti	5.80	3,703	730.6019	7,839.25	1,546.68
	Mpongwe	5.82	8,465	1,245.2015	17,982.20	2,645.18
	Mufulira	5.74	1,145	182.1695	2,398.89	381.66
	Ndola	5.74	1,035	154.1115	2,168.43	322.88

(continued)

Table 7 (continued)

Province	Districts	Annual solar irradiation	Total surface area	Total suitable area	Theoretical potential	Geographical potential
Total			**31,328**	**4,475.6557**	**65,726.77**	**9,390.02**
Western	Kalabo	5.86	18,065	4,243.4685	38,639.23	9,076.35
	Kaoma	5.85	22,099	1,082.8510	47,186,89	2,312.16
	Lukulu	5.89	15,589	3,287.7201	33,514.01	7,068.11
	Mongu	5.90	10,125	405.0000	21,804.19	872.17
	Senanga	5.93	15,205	1,449.0365	32,910.46	3,136.37
	Sesheke	5.91	29,423	2,050.7831	63,469.82	4,423.85
	Shang'ombo	5.92	15,880	1,602.2920	34,313.50	3,462.23
Total			**126,386**	**14,121.1512**	**271,908.65**	**30,380.45**

Table 8 Provincial solar energy theoretical and geographical potential

Province	Annual average solar irradiation	Total surface area	Total suitable area	Theoretical energy potential	Geographical energy potential
	(kWh/m²-day)	(km²)	(km²)	(TWh/year)	(TWh/year)
Lusaka	5.70	21,896.00	2,151.70	45,554.63	4,476.62
Luapula	5.78	50,567.00	6,066.92	106,760.30	12,808.87
Central	5.76	94,394.00	10,051.37	198,511.37	21,138.12
Copperbelt	5.75	31,328.00	4,475.66	65,726.77	9,390.02
Northern	5.83	147,186.00	13,017.14	313,025.37	27,683.98
Northwestern	5.74	125,826.00	11,515.06	263,421.22	24,107.19
Western	5.89	126,386.00	14,121.15	271,908.65	30,380.45
Southern	5.80	85,823.00	16,593.56	181,630.34	35,117.56
Eastern	5.68	69,208.00	45,72,0.38	143,482.03	9,478.75
Zambia	5.77	752,614.00	82,564.60	1,590,020.67	174,581.55

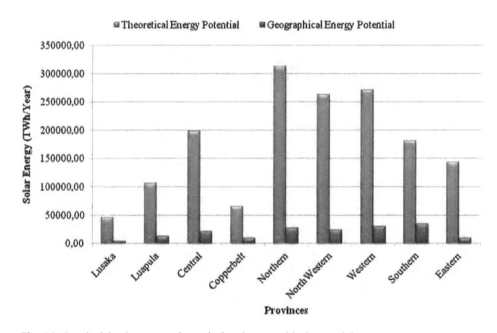

Fig. 16 Provincial solar energy theoretical and geographical potential

Table 12 and Fig.19 show that Southern Province, followed by Western Province have the highest potential while Lusaka province has the lowest potential for electricity generation from solar PV based technologies due to aforementioned issues. Figure 20 shows a comparison of the provincial theoretical and geographical solar energy potential and the technical solar electricity potential. It is worth noting that due to technical characteristic of the solar cell technologies and land requirements the technical solar electricity generation potential is lower as compared to the solar energy received on these potential sites. Hence, this presents the need to select

Table 9 District solar PV technical power potential by technology

Province	Districts	Technical power potential (GW)				
		mc-Si	pc-Si	a-Si	CIS	CdTe
Eastern	Chama	104.63	91.55	48.83	73.24	66.58
	Chipata	7.54	6.60	3.52	5.28	4.80
	Chadiza	8.57	7.50	4.00	6.00	5.45
	Petauke	70.45	61.64	32.88	49.31	44.83
	Katete	19.45	17.02	9.08	13.61	12.38
	Lundazi	15.16	13.26	7.07	10.61	9.65
	Nyimba	77.66	67.95	36.24	54.36	49.42
	Mambwe	23.12	20.23	10.79	16.19	14.71
Total	-	**326.57**	**285.75**	**152.40**	**228.60**	**207.82**
Lusaka	Chongwe	95.92	83.93	44.76	67.14	61.04
	Kafue	41.99	36.74	19.60	29.39	26.72
	Luangwa	14.51	12.69	6.77	10.16	9.23
	Lusaka	1.28	1.12	0.60	0.89	0.81
Total	-	**153.69**	**134.48**	**71.72**	**107.59**	**97.80**
Southern	Choma	92.83	81.23	43.32	64.98	59.08
	Gwembwe	34.15	29.88	15.94	23.90	21.73
	Itezhi-tezhi	277.04	242.41	129.28	193.93	176.30
	Kalomo	223.99	195.99	104.53	156.79	142.54
	Kazungula	312.11	273.10	145.65	218.48	198.62
	Livingstone	6.56	5.74	3.06	4.59	4.17
	Mazabuka	79.48	69.55	37.09	55.64	50.58
	Monze	48.42	42.37	22.60	33.90	30.81
	Namwala	19.22	16.82	8.97	13.46	12.23
	Siavonga	52.97	46.35	24.72	37.08	33.71
	Sinazongwe	38.48	33.67	17.96	26.94	24.49
Total	-	**1,185.25**	**1,037.10**	**553.12**	**829.68**	**754.25**
Luapula	Chienge	13.98	12.23	6.52	9.78	8.89
	Kawambwa	59.77	52.30	27.89	41.84	38.03
	Mansa	171.42	149.99	79.99	119.99	109.08
	Milenge	40.83	35.73	19.06	28.58	25.98
	Mwense	56.08	49.07	26.17	39.26	35.69
	Nchelenge	19.35	16.93	9.03	13.55	12.32
	Samfya	71.92	62.93	33.56	50.35	45.77
Total	-	**433.35**	**379.18**	**202.23**	**303.35**	**275.77**
North Western	Chavuma	149.39	130.72	69.72	104.58	95.07
	Kabompo	62.80	54.95	29.30	43.96	39.96
	Kasempa	67.44	59.01	31.47	47.21	42.92
	Mufumbwe	58.07	50.82	27.10	40.65	36.96
	Mwinilunga	64.48	56.42	30.09	45.14	41.03
	Solwezi	139.07	121.68	64.90	97.35	88.50
	Zambezi	281.25	246.09	131.25	196.87	178.98
Total	-	**822.50**	**719.69**	**383.84**	**575.75**	**523.41**

(continued)

Table 9 (continued)

Province	Districts	Technical power potential (GW)				
Province	Districts	mc-Si	pc-Si	a-Si	CIS	CdTe
Province	Districts	Technical power potential (GW)				
		Mc-Si	Pc-Si	a-Si	CIS	CdTe
Northern	Chinsali	43.96	38.47	20.52	30.77	27.98
	Chilubi	13.26	11.61	6.19	9.28	8.44
	Isoka	33.71	29.49	15.73	23.59	21.45
	Kaputa	148.24	129.71	69.18	103.77	94.34
	Kasama	82.15	71.88	38.34	57.50	52.28
	Luwingu	74.50	65.19	34.77	52.15	47.41
	Mbala	70.35	61.55	32.83	49.24	44.77
	Mpika	182.40	159.60	85.12	127.68	116.07
	Mporokoso	106.88	93.52	49.88	74.81	68.01
	Mpulungu	55.67	48.71	25.98	38.97	35.43
	Mungwi	60.09	52.58	28.04	42.06	38.24
	Nakonde	58.58	51.26	27.34	41.01	37.28
Total	-	**929.80**	**813.57**	**433.90**	**650.86**	**591.69**
Central	Chibombo	99.74	87.27	46.54	69.81	63.47
	Kabwe	6.77	5.93	3.16	4.74	4.31
	Kapiri-Mposhi	113.50	99.31	52.96	79.45	72.22
	Mkushi	129.84	113.61	60.59	90.88	82.62
	Mumbwa	164.56	143.99	76.80	115.19	104.72
	Serenje	203.55	178.11	94.99	142.49	129.53
Total	-	**717.96**	**628.21**	**335.05**	**502.57**	**456.88**
Copperbelt	Chililabombwe	10.05	8.79	4.69	7.04	6.40
	Chingola	17.60	15.40	8.21	12.32	11.20
	Kalulushi	12.81	11.21	5.98	8.97	8.15
	Kitwe	13.76	12.04	6.42	9.63	8.76
	Luanshya	11.01	9.63	5.14	7.70	7.00
	Lufwanyama	89.32	78.15	41.68	62.52	56.84
	Masaiti	52.19	45.66	24.35	36.53	33.21
	Mpongwe	88.94	77.83	41.51	62.26	56.60
	Mufulira	13.01	11.39	6.07	9.11	8.28
	Ndola	11.01	9.63	5.14	7.71	7.01
Total	-	**319.69**	**279.73**	**149.19**	**223.78**	**203.44**
Western	Kalabo	303.10	265.22	141.45	212.17	192.88
	Kaoma	77.35	67.68	36.10	54.14	49.22
	Lukulu	234.84	205.48	109.59	164.39	149.44
	Mongu	28.93	25.31	13.50	20.25	18.41
	Senanga	103.50	90.56	48.30	72.45	65.87
	Sesheke	146.48	128.17	68.36	102.54	93.22
	Shang'ombo	114.45	100.14	53.41	80.11	72.83
Total	-	**1,008.65**	**882.57**	**470.71**	**706.06**	**641.87**

Table 10 Provincial solar PV technical power potential per technology

Province	Technical Power Potential (GW)				
	mc-Si	pc-Si	a-Si	CIS	CdTe
Lusaka	153.69	134.48	71.72	107.59	97.80
Luapula	433.35	379.18	202.23	303.35	275.77
Central	717.96	628.21	335.05	502.57	456.88
Copperbelt	319.69	279.73	149.19	223.78	203.44
Northern	929.80	813.57	433.90	650.86	591.69
North-Western	822.50	719.69	383.84	575.75	523.41
Western	1,008.65	882.57	470.71	706.06	641.87
Southern	1,185.25	1,037.10	553.12	829.68	754.25
Eastern	326.57	285.75	152.40	228.60	207.82
Zambia	**5,897.46**	**5,160.28**	**2,752.16**	**4,128.24**	**3,752.93**

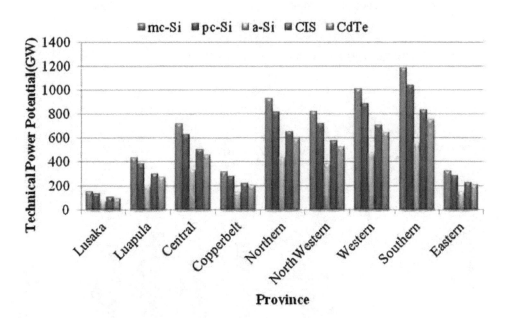

Fig. 17 Provincial solar PV technical power potential per technology

suitable solar cell technology for application in the solar energy harvesting systems for optimal solar energy utilization.

Figure 21 shows the comparison of solar PV technologies for electricity generation potential for Zambia considering the available suitable areas and the technology characteristics. İt is observed that monocrystalline provides the highest electricity generation potential followed by polycrystalline and least amorphous. This is mainly due to the differences in amount of land area requirements for the same peak power and the ability of the technology to convert the solar energy into electrical energy (efficiency).

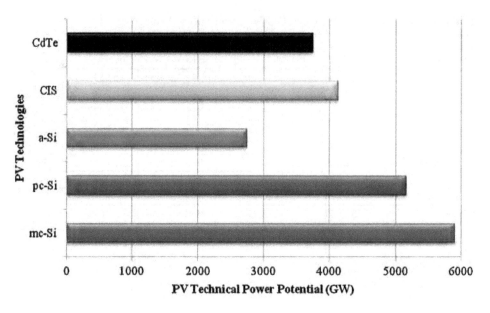

Fig. 18 Comparison of solar PV technical power potential per technology of Zambia

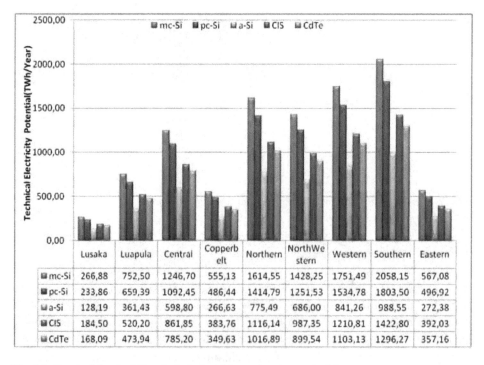

	Lusaka	Luapula	Central	Copperb elt	Northern	NorthWe stern	Western	Southern	Eastern
▨ mc-Si	266,88	752,50	1246,70	555,13	1614,55	1428,25	1751,49	2058,15	567,08
▨ pc-Si	233,86	659,39	1092,45	486,44	1414,79	1251,53	1534,78	1803,50	496,92
▨ a-Si	128,19	361,43	598,80	266,63	775,49	686,00	841,26	988,55	272,38
▨ CIS	184,50	520,20	861,85	383,76	1116,14	987,35	1210,81	1422,80	392,03
▨ CdTe	168,09	473,94	785,20	349,63	1016,89	899,54	1103,13	1296,27	357,16

Fig. 19 Provincial solar PV technical electricity generation potential

Table 11 District solar PV technical electricity generation potential by technology

Province	Districts sites	Technical electricity generation potential (TWh/Year) Solar PV technologies				
		mc-Si	pc-Si	a-Si	CIS	CdTe
Eastern	Chama	181.69	159.21	87.27	125.60	114.43
	Chipata	13.09	11.47	6.29	9.05	8.24
	Chadiza	14.88	13.04	7.15	10.28	9.37
	Petauke	122.33	107.19	58.76	84.57	77.05
	Katete	33.77	29.59	16.22	23.35	21.27
	Lundazi	26.32	23.07	12.64	18.20	16.58
	Nyimba	134.85	118.17	64.77	93.22	84.93
	Mambwe	40.15	35.18	19.29	27.76	25.29
Total	-	**567.08**	**496.92**	**272.38**	**392.03**	**357.16**
Lusaka	Chongwe	166.56	145.95	80.00	115.14	104.90
	Kafue	72.91	63.89	35.02	50.41	45.92
	Luangwa	25.19	22.07	12.10	17.41	15.87
	Lusaka	2.22	1.94	1.07	1.53	1.40
Total	-	**266.88**	**233.86**	**128.19**	**184.50**	**168.09**
Southern	Choma	161.20	141.25	77.43	111.44	101.53
	Gwembwe	59.30	51.96	28.48	40.99	37.35
	Itezhi-tezhi	481.06	421.54	231.06	332.56	302.99
	Kalomo	388.94	340.82	186.81	268.88	244.97
	Kazungula	541.97	474.91	260.31	374.67	341.35
	Livingstone	11.39	9.98	5.47	7.87	7.17
	Mazabuka	138.02	120.94	66.29	95.41	86.93
	Monze	84.08	73.68	40.39	58.13	52.96
	Namwala	33.38	29.25	16.03	23.08	21.03
	Siavonga	91.98	80.60	44.18	63.58	57.93
	Sinazongwe	66.83	58.56	32.10	46.20	42.09
Total	-	**2,058.15**	**1,803.50**	**988.55**	**1,422.80**	**1,296.27**
Luapula	Chienge	24.27	21.27	11.66	16.78	15.28
	Kawambwa	103.78	90.94	49.85	71.75	65.37
	Mansa	297.66	260.83	142.97	205.77	187.47
	Milenge	70.90	62.13	34.06	49.02	44.66
	Mwense	97.39	85.34	46.78	67.32	61.34
	Nchelenge	33.61	29.45	16.14	23.23	21.17
	Samfya	124.89	109.44	59.99	86.34	78.66
Total	-	**752.50**	**659.39**	**361.43**	**520.20**	**473.94**
Northwestern	Chavuma	259.42	227.32	124.60	179.34	163.39
	Kabompo	109.04	95.55	52.37	75.38	68.68
	Kasempa	117.11	102.62	56.25	80.96	73.76
	Mufumbwe	100.84	88.37	48.44	69.71	63.51
	Mwinilunga	111.97	98.12	53.78	77.40	70.52
	Solwezi	241.48	211.61	115.99	166.94	152.09
	Zambezi	488.38	427.95	234.57	337.62	307.59
Total	-	**1,428.25**	**1,251.53**	**686.00**	**987.35**	**899.54**

(continued)

Table 11 (continued)

Province	Districts sites	Technical electricity generation potential (TWh/Year) Solar PV technologies				
		mc-Si	pc-Si	a-Si	CIS	CdTe
Northern	Chinsali	76.34	66.90	36.67	52.77	48.08
	Chilubi	23.03	20.18	11.06	15.92	14.51
	Isoka	58.53	51.29	28.11	40.46	36.86
	Kaputa	257.42	225.57	123.64	177.96	162.13
	Kasama	142.65	125.00	68.51	98.61	89.84
	Luwingu	129.37	113.36	62.14	89.43	81.48
	Mbala	122.116	107.04	58.67	84.45	76.94
	Mpika	316.173	277.54	152.13	218.96	199.48
	Mporokoso	1851.59	162.63	89.14	128.30	116.89
	Mpulungu	96.168	84.71	46.43	66.83	60.89
	Mungwi	104.35	91.43	50.12	72.13	65.72
	Nakonde	101.72	89.13	48.86	70.32	64.07
Total	-	**1,614.55**	**1,414.79**	**775.49**	**1,116.14**	**1,016.89**
Province	Districts Sites	Technical electricity generation potential (TWh/year) Solar PV technologies				
		Mc-Si	Pc-Si	a-Si	CIS	CdTe
Central	Chibombo	173.19	151.76	83.18	119.72	109.08
	Kabwe	11.76	10.31	5.65	8.13	7.41
	Kapiri-Mposhi	197.08	172.70	94.66	136.24	124.13
	Mkushi	225.45	197.56	108.29	155.86	142.00
	Mumbwa	285.75	250.40	137.25	197.54	179.97
	Serenje	353.46	309.73	169.77	244.35	222.62
Total	-	**1,246.70**	**1,092.45**	**598.80**	**861.85**	**785.20**
Copperbelt	Chililabombwe	17.45	15.29	8.38	12.06	10.99
	Chingola	30.56	26.78	14.68	21.12	19.25
	Kalulushi	22.25	19.49	10.69	15.38	14.01
	Kitwe	23.89	20.94	11.48	16.52	15.05
	Luanshya	19.11	16.75	9.18	13.21	12.04
	Lufwanyama	155.09	135.90	74.49	107.22	97.68
	Masaiti	90.62	79.41	43.53	62.64	57.07
	Mpongwe	154.45	135.34	74.18	106.77	97.27
	Mufulira	22.60	19.80	10.85	15.62	14.23
	Ndola	19.11	16.75	9.18	13.21	12.04
Total	-	**555.13**	**486.44**	**266.63**	**383.76**	**349.63**
Western	Kalabo	526.33	461.21	252.80	363.85	331.50
	Kaoma	134.31	117.69	64.51	92.85	84.59
	Lukulu	407.79	357.33	195.86	281.90	256.83
	Mongu	50.23	44.02	24.13	34.73	31.64
	Senanga	179.73	157.49	86.33	124.25	113.20
	Sesheke	254.36	222.89	122.17	175.84	160.21
	Shang'ombo	198.74	174.15	95.46	137.39	125.17
	-	**1,751.49**	**1,534.78**	**841.26**	**1,210.81**	**1,103.13**
Zambia	-	**10,240.73**	**8,973.66**	**4,918.72**	**7,079.44**	**6,449.86**

Table 12 Provincial solar PV technical electricity generation potential by technology

| Provinces | Technical Electricity Generation Potential (TWh/Year) | | | | |
| | Solar PV technologies | | | | |
	mc-Si	pc-Si	a-Si	CIS	CdTe
Lusaka	266.88	233.86	128.19	184.50	168.09
Luapula	752.50	659.39	361.43	520.20	473.94
Central	1,246.70	1,092.45	598.80	861.85	785.20
Copperbelt	555.13	486.44	266.63	383.76	349.63
Northern	1,614.55	1,414.79	775.49	1,116.14	1,016.89
Northwestern	1,428.25	1,251.53	686.00	987.35	899.54
Western	1,751.49	1,534.78	841.26	1,210.81	1,103.13
Southern	2,058.15	1,803.50	988.55	1,422.80	1,296.27
Eastern	567.08	496.92	272.38	392.03	357.16
Zambia	**10,240.73**	**8,973.66**	**4,918.72**	**7,079.44**	**6,449.86**

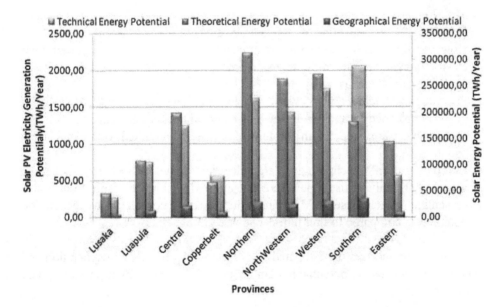

Fig. 20 Comparison of theoretical, geographical, and technical solar energy potential

While Zambia has abundance suitable areas (Fig. 14) and almost evenly distributed sunlight (Figs. 1 and 2) across the country, the focus on surface and suitable areas in the nine provinces and solar irradiation levels, the following can be identified. These factors however should be considered in the planning of national energy mix and also for management of electricity in the national grid once the penetration of solar PV technologies increases and becomes a significant part in the national electricity generation.

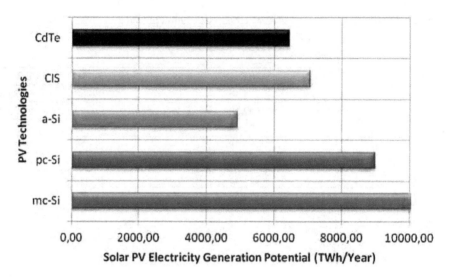

Fig. 21 Comparison of solar PV cell technologies electricity generation potential

- The highest theoretical solar energy potential is in Northern Province (313,025.37TWh/year) due to large surface areas of the province.
- However, the highest geographical and technical solar energy potential for solar electricity generation is in Southern Province (35,117.56TWh/year) due to large available suitable areas.
- From highest yield point of view, due to abundance of sunlight received by Western province ($5.89kWh/m^2$-day), the annual yields per installed solar PV systems peak are expected in Western province as compared to the rest of the country.
- Comparing the PV technologies, large electricity generation differences can be observed not only at district level but also at provincial levels. Table 13 indicates crystalline silicon based PV technologies have higher electricity generation potential as compared to thin film per square kilometer.

Table 13 summarizes the estimated solar PV technical electricity generation and solar PV power capacity potential in Zambia for each nine (9) provinces investigated in this chapter.

Conclusion

This chapter provides an approach for identifying and mapping the potential sites for sustainable development of solar PV technologies based power plants using GIS spatial analysis. The chapter has integrated the geographical and technological factors as well as the Laws of Zambia on environmental protection and pollution control legislative framework for evaluating the electricity generation potential and feasible sites suitable for sustainable PV systems deployment across Zambia.

Table 13 National solar PV technical electricity generation potential

Provinces	Annual average solar irradiation (kWh/m²·day)	Total surface area (km²)	Total suitable area (km²)	Theoretical energy potential (TWh/Year)	Geographical energy potential (TWh/year)	Technical Power Potential (GWp) Solar PV technologies					Technical electricity generation potential (TWh/Year) Solar PV technologies				
						mc-Si	pc-Si	a-Si	CIS	CdTe	mc-Si	pc-Si	a-Si	CIS	CdTe
Lusaka	5.70	21,896	2,152.70	45,554.63	4,476.62	153.69	134.48	71.72	107.59	97.80	266.88	233.86	128.19	184.50	168.09
Luapula	5.78	50,567	6,066.92	106,760.30	12,808.87	433.35	379.18	202.23	303.35	275.77	752.50	659.39	361.43	520.20	473.94
Central	5.76	94,394	10,051.37	198,511.37	21,138.12	717.96	628.21	335.05	502.57	456.88	1,246.70	1,092.45	598.80	861.85	785.20
Copperbelt	5.75	31,328	4,475.66	65,726.77	9,390.02	319.69	279.73	149.19	223.78	203.44	555.13	486.44	266.63	383.76	349.63
Northern	5.83	147,186	13,017.14	313,025.37	27,683.98	929.80	813.57	433.90	650.86	591.69	1,614.55	1,414.79	775.49	1,116.14	1,016.89
Northwestern	5.74	125,826	11,515.06	263,421.22	24,107.19	822.50	719.69	383.84	575.75	523.41	1,428.25	1,251.53	686.00	987.35	899.54
Western	5.89	126,386	14,121.15	271,908.65	30,380.45	1,008.65	882.57	470.71	706.06	641.87	1,751.49	1,534.78	841.26	1,210.81	1,103.13
Southern	5.80	85,823	16,593.56	181,630.34	35,117.56	1,185.25	1,037.10	553.12	829.68	754.25	2,058.15	1,803.50	988.55	1,422.80	1,296.27
Eastern	5.68	69,208	4,572.038	143,482.03	9,478.75	326.57	285.75	152.40	228.60	207.82	567.08	496.92	272.38	392.03	357.16
Zambia	**5.77**	**752,614**	**82,564.60**	**1,590,020.67**	**174,581.55**	**5,897.46**	**5,160.28**	**2,752.16**	**4,128.24**	**3,752.93**	**10,240.73**	**8,973.66**	**4,918.72**	**7,079.44**	**6,449.86**

Thus, this chapter shows that Zambia has vast available solar energy technical potential for PV electricity generation. The larger PV electricity generation potential variability at district and provincial level is highly linked with the local geographical features and terrain which affect the availability of suitable area and also local solar energy resource. Therefore, integration and generation of electricity from PV systems has greater potential to mitigate the current energy shortage and increase access to energy for all in Zambia. Furthermore, the suitable land areas in almost all districts and provinces is large enough for solar energy harvesting at utility-scale PV system capable of covering the present and future total electricity demands for Zambia. The identified potential sites have a total of available suitable area of 82,564.601 km^2 representing 10.97% of Zambia's total surface area equivalent to 5,897.46 GW technical power potential. This translates to 10,240.73TWh/year electricity generation potential considering annual average solar irradiation of 5.78 kWh/m^2-day and monocrystalline silicon solar PV technology mounted at optimal tilt angle. This potential has capacity to reduce CO_2 emission and contribute to achieve energy access for all and Sustainable Development Goals (SDGs).

The identification of potential sites and solar energy potential analysis will help improve the understanding of the potential solar energy, and PV technology can contribute to achieving sustainable national energy mix and increasing energy access for all in Zambia. Furthermore, it will help the government in setting up tangible energy targets and effective integration of solar PV systems into national energy mix. Hence, it is hoped that the suitability map established and the technical potential evaluated will help guide the decision makers and also the investors for planning future electricity generation targets and investment across the country and achieve the 2030 development goals.

References

Abdolvahhab Fetanat EK (2015) A novel hybrid MCDM approach for offshore wind farm site selection: a case study of Iran. Ocean Coast Manag 109:17–28. https://doi.org/10.1016/j.ocecoaman.2015.02.005

Addisu D, and Mekonnen PV (2015) A web-based participatory GIS (PGIS) for offshore wind farm suitability within Lake Erie, Ohio. Renew Sust Energ Rev 41:162–177. https://doi.org/10.1016/j.rser.2014.08.030, ScienceDirect

Ahmed Aly SS (2017) Solar power potential of Tanzania: identifying CSP and PV hot spots through a GIS multicriteria decision making analysis. Renew Energy 113:159–175. https://doi.org/10.1016/j.renene.2017.05.077, ScienceDirect

Ahmed HHA et al (2017) Energy performance, environmental impact, and cost assessment for a photovoltaic plant under Kuwait climate condition. Sustainable Energy technologies & Assessments 22:25–33

Alami Merrouni A, Mezrhab A, Mezrhab A (2014) CSP sites suitability analysis in the eastern region of Morocco. Energy Procedia 49:2270–2279. https://doi.org/10.1016/j.egypro.2014.03.240

Ali HM, Hojabri M, Hamada HM, Samsuri FB, Ahmed MN (2016) Performance evaluation of two PV technologies (c-Si and CIS) for building integrated photovoltaic based on tropical climate condition: a case study in Malaysia, energy and buildings. http://dxdoi.org/10.1016/j.enbuild.2016.03.052

Aly Sanoh AS (2014) The economics of clean energy resource development and grid interconnection in Africa. Renew Energy 62:599–609. https://doi.org/10.1016/j.renene.2013.09.017, ScienceDirect

Anthony Lopez BR (2012) U.S. renewable energy technical potentials: a GIS based analysis. NREL, U.S. Department of Energy. National Renewable Energy Laboratory (NREL, Colorado. NREL/TP-6A20-51946

Arthur Bossavy RG (2016) Sensitivity analysis in the technical potential assessment of onshore wind and ground solar photovoltaic power resources at regional scale. Appl Energy 182:145–153. https://doi.org/10.1016/j.apenergy.2016.08.075

Ayompe LD (2014) An assessment of the energy generation potential of photovoltaic systems in Cameroon using satellite-derived solar radiation datasets. Sustainable Energy Technologies and Assessments 7:257–264

Bowa CM (2017) Solar photovoltaic energy progress in Zambia: a review. SAUPEC 2017, 25th Southern African Universities Power Engineering Conference, 30 January-01 February, 2017. Stellenbosch, South Africa: SAUPEC 2017

Brewer JR (2014) Solar PV site suitability using GIS analytics to evaluate utility scale solar power potential in the U.S: South West Region. Department of Civil and Environmental Engineering. Brigham: Brigham Young University. Retrieved April 2017

British Standard (1998) Photovoltaic system performance monitoring, guidelines for measurement, data exchange and analysis:BS EN 61724:1998, IEC 61724: 1998. BSI 05.1999, British Standard

Chao-Rong Chen C-CH-J (2014) Ahybrid MCDM model for improving GIS-based solar farms site selection. (C.-S. Jwo, Ed.) International Journal of Photoenergy:1–9. https://doi.org/10.1155/2014/925370

Charabi Y, and Gastli A (2011) PV site suitability analysis using GIS-based spatial fuzzy multi criteria evaluation. Renew Energy 36:2554–2561. https://doi.org/10.1016/j.renene.2010.10.037, ScienceDirect

Chinairn (2013) Photovoltaic industry: abandoned projects on the plateau. Chinairn. Retrieved January 2016, from http://www.chinairn.com/print/3042250.html

Damon T, and Vasilis F (2011) Environmental impacts from the installation and operation of large-scale solar power plants. Renew Sust Energ Rev 15:3261–3270

Didler TG (2012) The RETScreen model for assessing potential of PV projects. RETScreen

England, N (2011) Solar Parks: maximising environmental benefits. Natural England

Fylladitakis E (2015) Environmental impacts of photovoltaic systems. Brunel University, London

Gauri SS (2013) Zambia renewable readiness assessment. International Renewable Energy Agency (IRENA), Lusaka

Geoffrey TK, and Tidwell VC (2013) Water use and supply concerns for utility-scale solar projects in the southwestern United States. Sandia National Laboratories, Livermore

Gipe P (1995) Wind energy comes of age. Wiley, New York

Interior Department, U (2010) Impacts of solar energy development and potential mitigation measures. I Draft solar programmatic environmental impact statement, vol 1, pp 1–302. N/A: Federal Register. Accessed in July 2017

IRENA (2012) Renewable energy technologies: cost analysis series volume1: power sector issue4/5. International Renewable Energy Agency (IRENA) working Paper

IRENA (2013) Zambia renewable readiness assessment 2013. International Renewable Energy Agency (IRENA), Lusaka

Ivan PR (2015) Analysis of insolation potential of Knjazevac Municipality (Serbia) using multi-criteria approach. Renewable and Sustainable Energy Review

Janke JR (2010) Multicriteria GIS modeling of wind and solar farms in Colorado. Renew Energy 35:2228–2234. https://doi.org/10.1016/j.renene.2010.03.014. ScienceDirect

Joss JW, and Watson MD (2015) Regional scale wind farm and solar farm suitability assessment using GIS-assisted multi-criteria evaluation. Landsc Urban Plan 138:20–31. https://doi.org/10.1016/j.landplan.2015.02.001. ScienceDirect

Kaoshan D et al (2015) Environmental issues associated with wind energy-a review. Renew Energy 75:911–921. https://doi.org/10.1016/j.renene.2014.10.074

Kynakis EK (2009) Performance analysis of a grid connected photovoltaic park on the island of Crete. Energy Convers Manag 50(3):433–438

Lopez AR (2012) U.S renewable energy technical potentials: a GIS-based analysis. Technical Report NREL/TP-6A20–5146, United States

Marcos Rodriques CM (2010) A method for the assessment of the visual impact caused by the large scale deployment of renewable energy facilities. Environ Impact Assess Rev 30:240–246. https://doi.org/10.1016/j.eiar.2009.10.004

Martin-Chivelet N (2016) Photovoltaic potential and land use estimation methodology. Energy 94: 233–242. https://doi.org/10.1016/j.energy.2015.10.108

Ming ZS (2015) Is the "Sun" still hot in China? The study of the present situation, problems and trends of the photovoltaic industry in China. Renew Sust Energ Rev 23:1224–1237

MMEWD (2008) National energy policy. Ministry of Mines, Energy and Water Development of Zambia, (MMEWD), Lusaka

Mwanza M, et al. (2016a) Assessment of solar energy source distribution and potential in Zambia. ISEM2016, 3rd international symposium on environment and morality, 4–5 November 2016. Alanya-Turkey: ISEM2016

Mwanza M, et al. (2016b) The potential of solar energy for sustainable water resource development and averting national social burden in rural areas of Zambia. ISEM2016, 3rd international symposium on environment and morality, 4–6 November 2016. Alanya-Turkey: ISEM2016

Nazli Yonca A (2010) GIS-based environmental assessment of wind energy systems for spatial planning: a case study from western Turkey. Renewable and Sustainable Energy Review 14:364–373. https://doi.org/10.1016/j.rser.2009.07.023. ScienceDirect

Quansah DA (2016) Solar photovoltaics in sub-Saharan Africa-addressing barriers, unlocking potential. Energy Economics Iberian Conference (Energy Procedia) 106:97–110. Lisbon, Portugal: ELSEVIER, ScienceDirect. https://doi.org/10.1016/j.egypro.2016.12.108

Robert CD (2014) Using the sun to decarbonize the power sector: the economic potential of photovoltaics and concentrating solar power. Appl Energy 135:704–720

Ronald CE (2016) GIS-based multi-criteria decision analysis in natural resource management. GIS-Based Multi-Criteria Decision Analysis. University of Tsukuba

Saidur R, Rahim NA, Islam MR, Solangi KH (2011) Environmental impact of wind energy. Renew Sust Energ Rev 15(5):2423–2430. Accessed in July 2017

Samuel AS, and Owusu PA (2016). A review of Ghana's energy sector national energy statistics and policy framework. Civil Environ Eng:1–27. Retrieved January 2017

Sanchez Lozano JM et al (2013) Geographical information system (GIS) and multi criteria decision making (MCDM) methods for the evaluation of solar farms locations: case study in south eastern Spain. Renewable and Sustainable Energy Review 24:544–556. https://doi.org/10.1016/j.rser.2013.03.019. ScienceDirect

Sanchez-Lozano JM, and García-Cascales MS (2014). Identification and selection of potential sites for onshore wind farms development in region of Murcia, Spain. Energy, 73(−), 311–324. https://doi.org/10.1016/j.energy.2014.06.024. ScienceDirect

SEFI/UNEP (2009) Global trends in sustainable energy investment (online). SEFI/UNEP. Retrieved from http://sefi.unep.org/english/globaltrends2009.html;

Shifeng W, and Sicong W (2015) Impact of wind energy on environment; a review. Renew Sust Energ Rev 49:437–443. https://doi.org/10.1016/j.rser.2015.04.137

Siheng W et al (2016) Selecting photovoltaic generation sites in Tibet using remote sensing and geographic analysis. Sol Energy 133:85–93. https://doi.org/10.1016/j.solener.2016.03.069. ScienceDirect

Siyasankari S, and Babu JSC (2015) Performance evaluation and validation of 5MW$_P$ grid connected solar photovoltaic plant in South India. Energy Convers Manag 100(2015):429–439, ScienceDirect

Suprava C. & Pradip K.S., (2015), Technical mapping of solar photovoltaic for the coal city of india, Renewables: Wind, Water, and Solar (2015) 2:11, DOI 10.1186/s40807-015-0013-1

Suri DQ (2005) PV-GIS: a web-based solar radiation database for the calculation of PV potential in Europe. International Journal of Sustainable Energy 24:55–67

The Environmental Council of Zambia (1994) Chapter 204 of the Laws of Zambia, the environmental protection and pollution control act. Ministry of Legal Affairs, Government of the Republic of Zambia, Lusaka

Tripathi B, Yadav P, Rathod S, Kumar M (2014) Performance analysis and comparison of two silicon material based photovoltaic technologies under actual climatic conditions in Western India. Energy Convers Manag 80:97–102

Tsoutsos T, et al (2005) Environmental impacts from the solar energy technologies. Energy Policy 33:289–296

Tsoutsos T et al (2009) Visual impact evaluation of a wind park in a Greek Island. Appl Energy 86: 546–553

Turlough G (2017) A case study identifying and mitigating the environmental and community impacts from construction of a utility-scale solar photovoltaic power plant in eastern Australia. Sol Energy 146:94–104

U.S. (2016). www.energy.gov. Retrieved November 2016, from Office of Energy Efficiency & Renewable Energy

Union of Concerned Scientists (2015) Environmental impacts of renewable energy technologies. Accessed in August, 2016 from Union of Concerned Scientists: Science for a healthy planet and safer world: www.ucsusa.org/clean-energy/renewable-energy/environmental-impacts#bf-toc-0

Uyan Mevlut (2013) GIS-based solar farms site selection using analytic hierarchy process in Karapinar region, Konya/Turkey. Energy. Renewable and Sustainable Energy Review 28:11–17. https://doi.org/10.1016/j.rser.2013.07.042. ScienceDirect

Walimwipi HS (2012) Investment incentives for renewable energy in southern Africa: case study of Zambia. International Institute for Sustainable Development (IISD), Lusaka

Wang C, and Prinn RG (2010) Potential climate impacts and reliability of a very large-scale wind farm. Atmospheric Chemical and Physical 10:2053–2061

Yan-wei SA (2013) GIS-based approach for potential analysis of solarPV generation at the regional scale: a case study of Fujian Province. Energy Policy 58:248–259

Yassine CA (2011) PV site suitability analysis using GIS-based spatial fuzzy multi-criteria evaluation. Renew Energy 36:2554–2561

Yassine C, and Adel G (2012) Spatio-temporal assessment of dust risk maps for solar energy system using proxy data. Renew Energy 44:23–31. https://doi.org/10.1016/j.renene.2011.12.005, ScienceDirect

10

Climate Change Adaptation in Southern Africa: Universalistic Science or Indigenous Knowledge or Hybrid

Tafadzwa Mutambisi, Nelson Chanza, Abraham R. Matamanda, Roseline Ncube and Innocent Chirisa

Contents

Introduction
Climate Change Response in Southern Africa
Knowledge Driving Climate Change Responses in Southern Africa
Vulnerability Concept
Ecosystem-Based Adaptation Concept
REDD+ Concept
Resilience Concept
Experiences with Climate Change Responses in Southern Africa

T. Mutambisi
Department of Demography Settlement and Development, University of Zimbabwe, Harare, Zimbabwe

N. Chanza
Department of Geography, Bindura University of Science Education, Bindura, Zimbabwe

A. R. Matamanda
Department of Urban and Regional Planning, University of the Free State, Bloemfontein, South Africa

R. Ncube
Faculty of Gender and Women Studies, Women's University in Africa, Harare, Zimbabwe

I. Chirisa (✉)
Department of Demography Settlement and Development, University of Zimbabwe, Harare, Zimbabwe

Department of Urban and Regional Planning, University of the Free State, Bloemfontein, South Africa

Towards a Hybrid Knowledge Approach in Climate Change Adaptation ....
Conclusions and Recommendations
References

Abstract

The aims of this chapter are to seek answer, through a document review, case studies, and thematic content analysis, to which direction Southern Africa should take in the face of climate change and to suggest a framework for adaptations by communities experiencing climatic events. Acknowledging that the fundamental set of ideas provided by indigenous knowledge (IK) works best at a small scale, the chapter argues for the need to seriously value IK-based response practices in the knowledge hybridization agenda. The worsening vulnerability potentiated by the increasing magnitude and severity of climate change impacts is a reminder that local-based indigenous response practices in Africa need to be complemented. Adaptation to climate change calls for real and surreal measures all being applied in combination. Across Africa, these measures have, at times, included the preservation of forest resources which increased carbon sinking and enhanced community resilience against climate change. Universalistic and ortho-dox sciences have punctuated and amplified these efforts by speaking of such initiatives as mitigation and adaptation through programs, e.g., Reducing Emis-sions from Deforestation and Forest Degradation (REDD+) and ecosystem-based adaptation (EbA). The merits of the two approaches have resulted in increasing call among scholars for the merging of these programs with IK. However, it remains to be fully understood how such a hybrid approach could be operationalized without treating the latter as an inferior element in climate science discourses.

Introduction

The complexity of climate change has attracted global attention and the various institutions worldwide to come up with adaptation and resilience methodologies and strategies (UNDP 2018). Throughout history, people and societies have successfully adjusted to and coped with climate change and extreme events. While climate change is a global issue, it is felt on a local scale (Silber 2013). This is because climate change affects many dimensions of development, including society, politics, science, economics, and moral and ethical questions (Henderson et al. 2017; Ogolla 2016). Local institutions are therefore at the frontline of adaptation as they attempt to factor in the local effects and how to mitigate these negative effects of climate change. This has resulted in the development of initiatives where climate change is factored into a variety of development plans on how to manage the increasingly extreme disasters and their associated risks, how to protect coastlines and deal with sea-level rise, how to best manage lands and forests, how to deal with and plan for reduced water availability, how to develop resilient crop varieties, and how to protect energy and public infrastructures (UNFCCC 2007; Carter et al. 2015; de Witt 2018).

Carbon dioxide, the heat-trapping greenhouse gas (GHG) that is the main contributor to global warming, has high longevity as it can remain in the atmosphere for hundreds of years (Dabasso et al. 2014). Even if the global community manages to establish strategies that can successfully reduce GHG emissions into the atmosphere, climate change and global warming will still be experienced and affect future generations. Thus, climate change continues to be a persistent global issue (Bauer and Scholz 2010). In Southern Africa, the effects of climate change have been experienced at a large scale as the region lacks adequate technology and proper management of its resources (Thinda et al. 2020). In addition, the combination of other various aspects of the region, such as its geographical exposure that has low levels of human development, weak institutions, high levels of inequality, and high dependence on agriculture, makes the African people vulnerable to the effects of climate change, especially those in the Southern African region as it is warming up at a faster rate than the global average (Kupika et al. 2019).

Climate change adaptation is a complex process that requires efforts from different stakeholders. The various problems that currently overwhelm Africa, for example, poor governance, lack of technology, and prevalence of poverty, make it difficult to establish the best approaches to achieve climate change adaptation. Basically, there are two available options that can be employed by the governments, namely, indigenous knowledge systems (IKS) and universalistic science. The former refers to the tested and proven knowledge that is used by local communities from their many years of observing and experiencing environmental phenomena, whereas the latter includes sophisticated techniques and approaches that are often associated with Western technologies and advanced way of mainstreaming climate change (Chirisa et al. 2018). These two approaches have their individual merits.

While some proponents view IKS as primitive and less reliable (e.g., Briggs 2005; Widdowson and Howard 2008), others assert that universalistic science is often too complicated and costly for African communities, which will result in their failure to maintain such a sophisticated technology (Wallner 2005; Simpson 2010; Walsh 2011). This causes ambiguity with regard to choosing the best approach to be employed to achieve climate change adaptation. In the same vein, this chapter proposes a hybrid method that considers the integration of both IKS and universalistic science. Thus, how Southern Africa can best adapt to climate change is investigated by weighing the strengths and limitations of the two knowledge systems in the context of climate change responses in the region. Through a document review, case studies, and thematic content analysis, the study determines the direction that Southern Africa should take in the face of climate change. The area of interest is shown in Fig. 1, which consists of Angola, Botswana, Eswatini, Lesotho, Malawi, Mozambique, Namibia, South Africa, Zambia, and Zimbabwe. It isolates specific themes to show how climate change responses have been pursued in the region. The chapter begins with a conceptual treatment of climate change adaptation in the frame of the distinctive knowledge forms (IKS and conventional knowledge systems) in the theory and practice of climate change responses. It then draws from existing scholars how these knowledge forms have characterized climate change mitigation and adaptation in Southern Africa. The discussion highlights the strengths

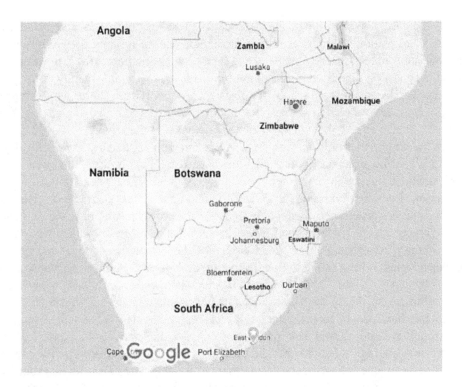

Fig. 1 The case study, Southern African Region.

and limitations of these knowledge systems and concludes that a hybrid system could be the best approach for driving an effective adaptation agenda in the region.

Climate Change Response in Southern Africa

Although climate change will place additional stress on Southern Africa's resources, some authors argue that there is considerable adaptive potential (Karsenty et al. 2012; Wang and Corson 2015; Kupika et al. 2019). In order to promote a meaningful and sustainable development in the region, the probable impacts of global climate change need to be considered, as well as the best alternative that fits the regional context. In the science of climate change, the two approaches that guide climate change responses are mitigation and adaptation. Mitigation is concerned with stabilizing the levels of heat-trapping GHG and reducing its emissions into the atmosphere, whereas the focus of adaptation is on taking measures to respond to the observed or anticipated climate change impacts (Klein et al. 2005; IPCC 2014). Climate change has been the greatest threat to provision of infrastructure and facilities for human improvement in countries in the Global South. Thus, it calls for the need to develop mitigation and adaptation strategies that enhance the resilience of the region (Mugambiwa 2018; Kupika et al. 2019). The major goal in mitigation is to avoid human interference in the climatic system and balances, whereas adaptation is a practice where human and natural systems devise appropriate measures in response to climate change impacts. In Southern Africa, most of the mechanisms developed by rural communities are complex and mainly based on culture and IK in the production of subsistence crops (Green 2008; Mugambiwa 2018). Mugambiwa (2018) stresses that Southern Africa's priority is to adapt to the climate change already being experienced anticipated in the region.

In the employment of the mitigation and adaptation approaches, it is critical to identify specific strategies, where IKS has been used, mainly in conjugation with universalistic science to strengthen the response system. The United Nations-based programs, Reducing Emissions from Deforestation and Forest Degradation (REDD +) and the ecosystem-based adaptation (EbA), are worth mentioning. These have been largely introduced through universal science that advocates for the reduction of GHG emissions into the atmosphere. Arguably, the knowledge of local and indigenous people has been found to be effective in the reduction of GHG emissions into the atmosphere (Kupika et al. 2019). Indigenous knowledge is already viewed as pivotal in several fields, such as sustainable development, agro-forestry, traditional medicine, biodiversity conservation, soil science, ethno-veterinary science, applied anthropology, and natural resource management (Mafongoya and Ajayi 2017). Many scholars highlight the significant role that IK plays in climate science and in facilitating climate change adaptation (Nyong et al. 2007; Green and Raygorodetsky 2010; Speranza et al. 2010; Crona et al. 2013; Ford et al. 2016; Chanza and Mafongoya 2017; Reyes-Garcia et al. 2019).

Knowledge Driving Climate Change Responses in Southern Africa

The different forms of knowledge, used either individually or jointly, characterize the climate change responses in Southern Africa to achieve climate change adaptation. The definition of IK is very complex and distinguishes sciences from other knowledge traditions (Mugambiwa 2018). There is no universal knowledge to which science is privy and to which all other knowledge traditions must defer. Knowledge is produced with relevance to specific contexts and questions, including Newtonian physics and Palikur astronomy (Perez et al. 2007). Indigenous knowledge is known as local or traditional knowledge and is defined as the beliefs and intellectual behaviors of indigenous societies, or local information about the relationship between humans and their environment (Mafongoya and Ajayi 2017). This knowledge is accumulated, developed, and transmitted through local community experiences and know-how across generations. Therefore, one can argue that IK is based on the cultural beliefs of a community and that it is orally transferred from one generation to another (Casper 2007; Bryan et al. 2011; Baudron 2011). Thus, there exists a binary divide between universalistic science and IK (Appignanesi 2018). Western science is seen to be systematic, objective, open, and dependent on a detached center of rationality and intelligence. From a radical perspective, universalistic science is understood to represent modernity, whereas IK is perceived as a traditional and backward way of life (Briggs 2005; Bauer and Scholz 2010). Clearly, these knowledge forms seem to be divergent and have been differently criticized. The cleavage between Western science and indigenous science is explored by Briggs (2005). Briggs (2005) argues that the use of IK is riddled with problems and challenges as it focuses on the (arte)factual, stating that this knowledge form has been romanticized. Wallner (2005) contends that Western science can become "very immoral" and "very dangerous" as it lacks subjectivity. From an African perspective, however, several scholars argue that IKS has been a sustainable factor in development of physical artefacts in space like infrastructure provisioning as it has minimized damage to the environment by allowing people to live in harmony with nature for generations (Kupika et al. 2019). Table 1 compares IK and universalistic science.

Local knowledge about ecosystems has been used in different parts of the world in an effort to respond to climate change. Universalism is defined as the epistemological and philosophical orientations taken by scientists of claims about the world or a particular issue in question (Green 2008). Western science is centralized and associated with the machinery of state, and its bearers believe in its superiority. The differences between IK and Western scientific knowledge are best described in the following grounds:

- Substantive grounds: there are differences in the subject matter between IK and Western knowledge.
- Methodological and epistemological grounds: the forms of knowledge employ different methods to investigate reality.
- Contextual grounds: IK is more deeply rooted in its environment (Mugambiwa 2018).

Therefore, it is undeniable that both Western science and IKS play significant roles in promoting sustainable development through mitigation and adaptation

Table 1 Major differences between IK and universalistic science

Major differences	Western/formal science	IK
Mode of transmission	Written, formally documented	Oral, repetitive
Substantive differences	To construct general explanations; is removed from people's daily lives	Concerned with immediate and concrete necessities of people's daily livelihoods
Methodological and epistemological differences	It is open, systematic, objective, and analytical. It advances by building rigorously on prior achievements	It is closed, non-systematic, and holistic rather than analytical. It advances on the basis of new experiences, not on the basis of deductive logic
Contextual differences	It is divorced from epistemic framework in search for universal validity	It exists in a local context anchored to a particular social group, in a particular setting, at a particular time

Adopted from: Mafongoya and Ajayi (2017)

measures. The full conceptualization of IK can greatly contribute to the strengthening of the resilience of poor societies or marginal areas in Southern Africa. One of the best approaches to employ for climate change adaptation in poor regions is the ecosystem-based approach (Chanza and de Wit 2016; SSSI 2018; USAID 2018). In Zimbabwe, this is crucial especially to small-scale farmers as they are more prone to the effects of climate change. It is predicted that billions of people, particularly those in developing countries, face water and food shortage as well as greater risks to health and life as a result of climate change (Shava et al. 2009; Shames et al. 2012; Wang and Corson 2015). Thus, fully assessing the benefits of both science and IK in the face of climate change in Southern Africa is crucial. In an effort to adapt to climate change, there are conceptual issues that need to be considered. This chapter considers four factors that may possibly provide insights into the adaptation approach that may be employed in Southern Africa. The conceptual issues of vulnerability, EbA, REDD+, and resilience are discussed in the following sections.

Vulnerability Concept

In relation to climate change, vulnerability is the degree to which a system is susceptible to, and unable to cope with, the adverse effects of climate change, including climate variability and extremes (UNFCCC 2007; IPCC 2014; Mavhura 2019). Climatic stress is already causing much pressure on the African region, making it highly vulnerable to the impacts of climate change (UNFCCC 2007; Bauer and Scholz 2010; Mugambiwa 2018). In Southern Africa, the poor predominantly depends on natural resource-based livelihoods, which are climate-sensitive. This makes them mostly vulnerable to poverty as they have minimum capacity to economically manage and cope with the damage caused by natural disasters or incremental degradation (Swinkels et al. 2019). Africa is well known to have diverse climates that are considered to be the most variable in the world on seasonal and decadal time scales. This makes the region's societies and ecosystems most prone to climatic events largely in the form of frequent floods and droughts (UNDP 2018). Africa's vulnerability to climate change also opens up room for its exposure to famine and widespread disruption of socioeconomic well-being as well as climate-sensitive diseases, such as diarrhea, malaria, and tuberculosis.

Ecosystem-Based Adaptation Concept

Adaptation to climate change is associated with adjusting to expected or actual climate. The goal of adaptation is to reduce the vulnerability of the environment, regions, or societies to the harmful and dire effects of climate change (Green 2008). It also involves making the most of any potential beneficial opportunities associated with climate change. Ecosystem-based adaptation therefore refers to the management of ecosystems to enhance ecological structures and functions essential for ecosystems and the people to adapt to multiple changes, including climate change

(Kupika et al. 2019). This approach has the potential to contribute multiple benefits through the reduction of vulnerability to climate change as well as contribute to socioeconomic development and conservation (Mafongoya and Ajayi 2017) Eco-system-based approaches include the maintenance of safety nets as a mechanism that best serves as a coping strategy during scarcity periods to enhance livelihood security. Some approaches implored under EbA include the following:

- Restoration of ecosystems that can reduce the exposure of human settlements to extreme weather events or climate change, e.g., combining the "building with nature" techniques with hard-engineering infrastructure to restore mangrove coastlines that reduce the risks of flood, erosion, and saline intrusion and support adaptation to sea-level rise (Bauer and Scholz 2010).
- Integration of sustainable ecosystem management practices into broad landscape-level planning processes. For example, integrated watershed management in peri-urban areas, which has proven to enhance water regulation to support the supply of water for drinking and hydroelectricity generation in cities.
- Payment for ecosystem services to diversify income generation and build adaptive capacity, for example, conserving a shoreline mangrove and coral reef system to maintain economies based on ecotourism, recreational activities, and fisheries.
- Climate mitigation by maintaining or enhancing carbon stocks with safeguards in place to support adaptation, for example, collective management of forested landscapes that promotes social learning to conserve forest function and structure, biodiversity and habitat connectivity, and climate-smart agriculture with agroforestry systems.

At the center of this approach lies the recognition of nonlinear feedbacks between social and ecological processes. Ecosystem-based approaches can be applied to all types of ecosystems and at different scales from local to global (NASA's Global Climate Change n.d.). Chanza and de Wit (2016) suggest that EbA should be best articulated at the local level where it can promote adaptation governance through IK. They also argue that the concept provides multiple benefits to society and the environment as it contributes to reducing vulnerability and increasing resilience to both climate and non-climate risks. Many cases in Africa have also shown that EbA approaches involving local people using their IKS can promote climate proofing and emission reduction (Hachileka 2010; Lo 2016; van Niekerk et al. 2017).

REDD+ Concept

The UN-based REDD+ program is an incentive-based tool that bases its work on public and private agents' self-interest in conservation strategies and the fact that they are well capable of calculating the opportunities as well as the costs associated with the reduction of GHG emissions into the atmosphere (Karsenty et al. 2012). The main concept of REDD+ is to make performance-based payments, which means that forest users and owners are paid to reduce GHGs emitted into the atmosphere

(Angelsen 2009). Payments for environmental services gives the forest owners the opportunity to receive incentives to motivate them to manage forests better and also minimize the rate at which deforestation occurs (Baudron 2011). Therefore, the concept of REDD+ is a rewarding mechanism to actors involved in keeping and restoring forests as well as a mechanism with the main objective of reducing carbon emissions (Karsenty et al. 2012; Skutsch and Ba 2010).

The concept of REDD+ has been implemented at a global level through the United Nations Development Program (UNDP) as a strategy to reduce GHG emissions into the atmosphere while promoting economic development, especially in developing regions, such as Africa (UNDP 2018). Developing countries have a competitive advantage in the carbon market if they choose to conserve their forests rather than convert them for agricultural use (Gantsho and Karani 2007). The REDD + mechanism thus invests in providing incentives to projects that promote forest conservation and conversion, afforestation, and reforestation (Wang and Corson 2015; Leach and Scoones 2013; Shames et al. 2012). This has led to the introduction of carbon markets that have successfully paved way for developing countries to earn funds through forest conservation. The plus sign on REDD+ indicates how the concept is also dedicated to the enhancement of forest carbon stock which can also be referred to as forest rehabilitation and regeneration, carbon removal, negative emissions, negative degradation, or carbon uptake (Angelsen 2009). Chanza and de Wit (2016) reveal opportunities for the success of this initiative if local communities can use their IKS in forest conservation. In Kenya and Tanzania, for example, the success of REDD+ projects has been attributed to participatory planning where the knowledge of local citizens in technical analysis has helped in addressing the drivers of deforestation and forest degradation (Richards and Swan 2014).

Resilience Concept

There is no universally accepted definition of resilience. The vast and growing literature available on urban resilience demonstrates the complexity of the concept as a target, as well as the challenges of mainstreaming recommendations into the urban development practice (UN-HABITAT 2017). The concept of resilience originated from ecological studies, exploring the varied ability of ecosystems to absorb and adapt to external pressures (Dau Kuir-Ayius 2016). The Intergovernmental Panel on Climate Change (IPCC) (2014, p. 1772) defines it as "the capacity of social, economic, and environmental systems to cope with a hazardous event or trend or disturbance, responding or reorganising in ways that maintain their essential function, identity, and structure, while also maintaining the capacity for adaptation, learning, and transformation." Understanding that human systems function as complex, interdependent, and integrated social-ecological systems is crucial to understanding how resilience-based planning, development, and management can protect life and assets as well as maintain the continuity of functions through any plausible shock or stress (Sekar et al. 2019; UN-HABITAT 2017; Dau Kuir-Ayius 2016). Thus, the Southern African region needs to develop resilient strategies that will assist them in coping with climate change (Doyon 2016; Dau Kuir-Ayius 2016; UN-HABITAT 2017; Kupika et al. 2019). Scholarship treating IK and climate change resilience is still developing (Bohensky and Maru 2011; DeAngelis 2013; Makate 2020). Despite the phenomenal growth in resilience literature, Bohensky and Maru (2011) express the lack of clarity and empirical evidence for the relationship between IK and resilience. Citing the utility of IK in agricultural projects, Makate (2020) argues that in establishing resilience to climate change, local communities should be seen as equal partners in the development process. The next section discusses the strengths and limitations of the existing climate change response practices in Southern African countries.

Experiences with Climate Change Responses in Southern Africa

The study found out that in its trajectory to curb the effects of climate change, Southern Africa is largely driven by modernity. However, due to the slower development of this region compared with the Western and European regions, it is difficult to implement technologically advanced measures owing to its limited manpower, poor resource management, and inadequate finances to make use of the science. Therefore, traditional knowledge and IK play a significant role in climate change adaptation in many rural communities (Green 2008; Mafongoya and Ajayi 2017; Kupika et al. 2019). However, due to modernity, it is now rare to find pure IKS practices that are not contaminated with the conventional construct of adaptation practices. Thus, it is important to fully understand indigenous-based practices that continue to shape the adaptation landscape in the African context and to assess the strengths and limitations of such strategies. This chapter argues that one way to complement the strengths of adaptation practices in the region is to complement the two forms of knowledge. Conscious of the blending gaps that have been highlighted in adaptation literature, the next sections also propose how the two knowledge systems can jointly work best.

In Southern Africa, there exists a diverse and unique group of cultures, including the Shona, Ndebele, Zulu, Tswana, Xhosa, Pedi, Tshangni, Venda, and Suthu, among many others (Mabogunje n.d.; Hanyani-Mlambo 2002). The main common aspect of all these African cultures is that they all started as hunters, gatherers, nomads, and pastoralists. Their traditions and way of living are still being practiced today, especially in marginal areas, for instance, in the Mbire District in Zimbabwe and the Khoisan in the Kalahari Desert of Botswana (Dube et al. 2014). In this regard, it can be safe to agree that traditional knowledge is also an ecological knowledge. It involves adaptation processes that have been handed down from generation to generation, and thus, they are useful in climate change adaptation (Dabasso et al. 2014).

The renewal of IK throughout generations has ensured the well-being of many Southern African people through ensuring food security, early warning systems, environmental conservation, and disaster risk management (Mawere 2010). Therefore, it can be deduced that the people rely on this traditional knowledge for social capital and food production as a way to ensure survival. The advent of universalistic science that has enabled the invention of technologies, machines, and concepts that are mostly unsustainable has led to IK being disregarded as it is considered as backward and old (Carmody 1998; Mugambiwa 2018).

Towards a Hybrid Knowledge Approach in Climate Change Adaptation

Climate change adaptation calls for more than just local experiences or scientific knowledge. Effective adaptation to climate change in Southern Africa requires the best knowledge regardless of where it is coming from (Shava et al. 2009). Hybrid knowledge from both sources is only possible through collaborative process, community participation, and involvement and interactions between the locals and scientists. It is also important to acknowledge that IK can provide solutions to food insecurity and poverty (Hussein and Suttie 2016). Appropriate and effective use of IK can promote a number of advantages that can match those of conventional science. These include environmental conservation and management of disasters in terms of prevention, mitigation, recovery, prediction or early warning, preparedness, and healing through traditional medicinal practices by producing traditional foods. In Kenya, the Maasai traders have relied on the benefits provided by the earth's resource, which has ensured a stable livelihood for centuries (Wang and Corson 2015).

As the world warms, traditional weather indicators may become less and less valuable. Individual species will adapt to the impacts of local climate change in idiosyncratic and unpredictable ways. Animals may change their behaviors or their ranges, whereas plants may begin flowering at different times. It is feared that these changes might render traditional knowledge less reliable. One way to strengthen the hybridization of IK and universalistic science in terms of climate change response is to revitalize or strengthen local institutions (Mararike 2011; Makate 2020). According to Makate (2020), the integration approach can work best only where climate adaptation practitioners build from existing local-based knowledge forms, rather than moving to replace them. When strengthened, local institutions can enhance the employment and scaling success of climate adaptation projects and innovations. Such efforts will improve information sharing, resource mobilization, stakeholder coordination, network establishment, and capacity building with local citizens as well as provide leadership and control of climate adaptation programs. The technological orientation of adaptation practice as seen through irrigation infrastructure in the agricultural sector, for instance, can be enhanced if local farmers are given sovereign rights over the traditional crop or animal varieties of their choice. Accordingly, it is reasonable to suggest that using the two knowledge forms individually will not provide solutions to the increasing threats of climate change in Southern Africa.

Conclusions and Recommendations

Given the sensitivity of the entire socioeconomic sectors and natural systems to the increasing climatic events, the Southern African region is expected to devise appropriate mitigation and adaptation practices. The region faces a quandary in terms of the trajectory that it should take in its development plan in the context of the existing and anticipated climate change threats. Climate change is starting to be factored into a variety of development plans on how to manage the increasingly extreme disasters faced by humanity and their associated risks, how to protect coastlines and deal with sea level rise, how to best manage land and forests, how to deal with and plan for reduced water availability, how to develop resilient crop varieties, and how to protect energy and public infrastructure. In other parts of the world, countries are working on building flood defenses, plan for heat waves and high temperatures, install water-permeable pavements to better deal with floods and stormwater, and improve water storage and use. However, in the absence of national or international climate policy direction, cities and local communities worldwide have focused on solving their own problems on climate change. The importance that communities in Southern Africa place on IKS in the face of climate change deserves recognition. However, due to changing seasons, IKS needs to be integrated with scientific methods as it cannot address the magnitude of the challenge alone. In the face of global climate change and its emerging challenges, unknowns, and uncertainties, it is important to base the decision-making for policies and actions on the best available knowledge. The multisectoral and cross-scale nature of climate change responses, such as REDD+ and ecosystem-based adaptation approaches, requires the integration of a range of disciplines, actors, and institutions interacting at different levels and influencing diverse decision networks.

References

Angelsen A (ed) (2009) Realising REDD+ National strategy and policy options. CIFOR, Bogor. Available online: http://www.cifor.cgiar.org

Appignanesi L (2018) Blurred binary code for the sustainable development of functional systems: blurred binary code. Syst Res Behav Sci 35(4):386–398. https://doi.org/10.1002/sres.2537

Baudron F (2011) Agricultural intensification – saving space for wildlife? PHD thesis, Wageningen University

Bauer S, Scholz I (2010) Adaptation to climate change in Southern Africa: new boundaries for sustainable development? Clim Dev 2(2):83–93. https://doi.org/10.3763/cdev.2010.0040

Bohensky EL, Maru Y (2011) Indigenous knowledge, science, and resilience: what have we learned from a decade of international literature on "integration"? Ecol Soc 16(4):6. https://doi.org/10.5751/ES-04342-160406

Briggs J (2005) The use of indigenous knowledge in development: problems and challenges. Progr Dev Stud 5(2):99–114

Briggs J, Moyo B (2012) The resilience of indigenous knowledge in small-scale African agriculture: key drivers. Scottish Geogr J 128(1):64–80. https://doi.org/10.1080/14702541.2012.694703

Bryan E, Akpalu W, Yesuf M et al (2011) Global Carbon Markets Opportunities for Sub Saharan Africa in Agriculture and Forestry. Clim Dev 2(4):309–330. https://doi.org/10.3763/cdev.2010.0057

Caputo S (n.d.) Urban resilience: a theoretical and empirical investigation, p 181

Carmody P (1998) Neoclassical practice and the collapse of industry in Zimbabwe: the cases of textiles, clothing, and footwear. Econ Geogr 74(4):319. https://doi.org/10.2307/144328

Carter JG, Cavan G, Connelly A, Guy S, Handley J, Kazmierczak A (2015) Climate change and the city: building capacity for urban adaptation. Progr Plann 95:1–66

Casper JK (2007) Agriculture: the food we grow and animals we raise. Chelsea House, New York

Chanza N, de Wit A (2016) Enhancing climate governance through indigenous knowledge: case in sustainability science. S Afr J Sci 112(3/4). https://doi.org/10.17159/sajs.2016/20140286

Chanza N, Mafongoya PL (2017) Indigenous-based climate science from the Zimbabwean experience: from impact identification, mitigation and adaptation. In: Mafongoya PL, Ajayi OC (eds) Indigenous knowledge systems and climate change management in Africa. Technical Centre for Agricultural and Rural Cooperation (CTA), Wageningen, 316pp

Chirisa I, Matamanda A, Mutambwa J (2018) Africa's dilemmas in climate change communication: universalistic science versus indigenous technical knowledge. In: Leal Filho W, Manolas E, Azul A, Azeiteiro U, McGhie H (eds) Handbook of climate change communication: vol. 1. Climate change management. Springer, Cham

Crona B, Wutich A, Slade A, Gartin M (2013) Perceptions of climate change: linking local and global perceptions through a cultural knowledge approach. Clim Change 119(2):519–531. https://doi.org/10.1007/s10584-013-0708-5

Dabasso BH, Taddese Z, Hoag D (2014) Carbon stocks in semi-arid pastoral ecosystems of northern Kenya. Pastoralism 4(1):5

Dau Kuir-Ayius D (2016) Building community resilience in mine impacted communities: a study on delivery of health services in Papua new guinea. Massy University. Available online: https://mro.massey.ac.nz/bitstream/handle/10179/9882/02_whole.pdf?sequence=2&isAllowed=y. Accessed 3 Nov 2019

De la Vega I, Puente JM, Sanchez M (2019) The collapse of Venezuela vs. the sustainable development of selected South American Countries. MDPI

de Witt S (2018) Measuring our investment in the future. Afr Eval J 6(2):a343. https://doi.org/10.4102/aej.v6i2.343

DeAngelis K (2013) Building resilience to climate change through indigenous knowledge: the case of Bolivia. Climate and Development Knowledge Network (CDKN). Available online: https://cdkn.org/wp-content/uploads/2013/03/Bolivia_InsideStory.pdf. Accessed 30 June 2010

Doyon A (2016) An investigation into planning for urban resilience through niche interventions. Doctoral, The University of Melbourne

Dube F, Nhapi I, Murwira A, et al (2014) Potential of weight of evidence modelling for gully erosion hazard assessment in Mbire District – Zimbabwe, pp 1–31

Ford JD, Cameron L, Rubis J, Maillet M, Nakashima D, Willox AC, Pearce T (2016) Including indigenous knowledge and experience in IPCC assessment reports. Nat Clim Change 6:349–353

Gantsho M, Karani P (2007) Entrepreneurship and innovation in development finance institutions for promoting the clean development mechanism in Africa. Dev South Afr 24(2):335–344

Green LJ (2008) 'Indigenous knowledge' and 'science': reframing the debate on knowledge diversity. Archaeologies 4(1):144–163

Green D, Raygorodetsky G (2010) Indigenous knowledge of a changing climate. Clim Change 100:239–242

Gujba H, Thorne S, Mulugetta Y, Rai K, Sokona Y (2012) Financing low carbon energy access in Africa. Energy Policy 47:71–78

Hachileka E (2010) Climate change adaptation strategies in the Chiawa community of the lower Zambezi game management area, Zambia. In: Andrade PA, Herrera FB, Cazzolla GR (eds) Building resilience to climate change: ecosystem-based adaptation and lessons from the field. IUCN, Gland, pp 89–98

Hanyani-Mlambo BT (2002) Strengthening the pluralistic agricultural extension system: a Zimbabwean case study. Food and Agriculture Organization of the United Nations (FAO), Rome

Henderson JV, Storeygard A, Deichmann U (2017) Has climate change driven urbanization in Africa? J Dev Econ 124:60–82

Hussein K, Suttie D (2016) Rural-urban linkages and food systems in sub-Saharan Africa: the rural dimension, vol 5. IFAD, Rome

IPCC (2014) Climate change 2014: impacts, adaptation, and vulnerability. Part B: regional aspects. In: Barros VR, Field CB, Dokken DJ, Mastrandrea MD, Mach KJ, Bilir TE, Chatterjee M, Ebi KL, Estrada YO, Genova RC, Girma B, Kissel ES, Levy AN, MacCracken S, Mastrandrea PR, White LL (eds) Contribution of working group II to the fifth assessment report of the Intergovernmental Panel on Climate Change. Cambridge University Press, Cambridge, UK/New York, p 688

Islam N, Winkel J (2017) Climate change and social inequality. DESA working paper no 152 ST/ESA/2017/DWP/152. Department of Economic & Social Affairs. Available online https://www.un.org/esa/desa/papers/2017/wp152_2017.pdf. Accessed 5 Mar 2020

Kahn ME, Walsh R (2014) Cities and the environment. National Bureau of Economic Research working paper 20503, vol 100, Cambridge, Massachusetts, United States

Karsenty A, Vogue A, Castell F (2012) "Carbon rights", REDD+ and payments for environmental services. Available online: https://doi.org/10.1016/j.envsci.2012.08.013

Klein RJ, Schipper ELF, Dessai S (2005) Integrating mitigation and adaptation into climate and development policy: three research questions. Environmental science & policy 8(6):579–588

Kupika OL, Gandiwa E, Godwell N, et al (2019) Local ecological knowledge on climate change and ecosystem-based adaptation strategies promote resilience in the middle Zambezi biosphere reserve, Zimbabwe, p 15

Leach M, Scoones I (2013) Carbon forestry in West Africa: the politics of models, measures and verification processes. Inst Dev Stud 23(5):957–967

Lo V (2016) Synthesis report on experiences with ecosystem-based approaches to climate change adaptation and disaster risk reduction. Technical series no 85. Secretariat of the Convention on Biological Diversity, Montreal, 106 p

Mabogunje AL (n.d.) Urban planning and the post-colonial state in Africa: a research overview. Available online: https://www.jstor.org/stable/pdf/524471.pdf?refreqid=excelsior%3Acebfc2f9b1cef3793196abe2ba7234cd. Accessed 11 Feb 2020

Mafongoya PL, Ajayi OC (eds) (2017) Indigenous knowledge systems and climate change management in Africa. CTA, Wageningen

Makate C (2020) Local institutions and indigenous knowledge in adoption and scaling of climate-smart agricultural innovations among sub-Saharan smallholder farmers. Int J Clim Change Strat Manag 12(2):270–287

Mararike CG (2011) Survival strategies in rural Zimbabwe: the role of asset, indigenous knowledge and organisation. Best Practices Books, Harare

Mavhura E (2019) Systems analysis of vulnerability to hydrometeorological threats: an exploratory study of vulnerability drivers in Northern Zimbabwe. Springer. Available online: https://doi.org/10.1007/s13753-019-0217-x

Mawere M (2010) Indigenous knowledge systems' (IKSs) potential for establishing a moral, virtuous society: lessons from selected IKSs in Zimbabwe and Mozambique. J Sustain Dev Afr 12(7):209–221

Moutinho P, Schwartzman S (eds) (2005) Tropical deforestation and climate change. Brasilia: Instituto de Pesquisa Ambiental da Amazônia and Environmental Defense

Mugambiwa SS (2018) Adaptation measures to sustain indigenous practices and the use of indigenous knowledge systems to adapt to climate change in Mutoko rural district of Zimbabwe. Jàmbá J Disaster Risk Stud 10(1):1–9

NGCC (n.d.) Climate change adaptation and mitigation. Available online https://climate.nasa.gov/solutions/adaptation-mitigation. Accessed 5 Mar 2020

Nyong A, Adesina F, Elasha BO (2007) The value of indigenous knowledge in climate change mitigation and adaptation strategies in the African Sahel. Mitigat Adapt Strat Global Change 12:787–797

Ogolla PA (2016) Africa and the plight of climate change. Development 2016(59):373–376

P Perez, C., Roncoli, C., Neely, C., & Steiner, J. L. (2007). Can carbon sequestration markets benefit low-income producers in semi-arid Africa? Potentials and challenges. Agric Syst, 94(1), 2-12.

Read "Advancing the Science of Climate Change" at NAP.edu (n.d.). https://doi.org/10.17226/12782.

Reyes-García V, García-del-Amo D, Benyei P, Fernández-Llamazares A, Gravani K, Junqueira AB, Labeyrie V, Li X, Matias DMS, McAlvay A, Mortyn PG, Porcuna-Ferrer A, Schlingmann A, Soleymani-Fard R (2019) A collaborative approach to bring insights from local observations of climate change impacts into global climate change research. Curr Opin Environ Sustain 39:1–8. https://doi.org/10.1016/j.cosust.2019.04.007

Richards M, Swan SR (2014) Participatory subnational planning for REDD+ and other land use programmes: methodology and step-by-step guidance. SNV Netherlands Development Organisation, REDD+ Programme, Ho Chi Minh

SDG Lead (2018) Final evaluation of the "REDD+ Governance and Finance Integrity for Africa" programme

Sekar S, Lundin K, Tucker C, et al (2019) Building resilience a green growth framework for mobilizing mining investment. World Bank Publications, Washington, DC. Available online: http://documents.worldbank.org/curated/en/689241556650241927/pdf/Building-Resilience-A-Green-Growth-Framework-for-Mobilizing-Mining-Investment.pdf. Downloaded 6 Nov 2019

Shames S, Wollenberg E, Buck LE, et al (2012) Institutional innovations in African smallholder carbon projects. CGIAR Research Program on Climate Change, Agriculture and Food Security (CCAFS) 8. CCAFs, Copenhagen. Available online http://www.ccafs.cgiar.org

Shava S, O'Donoghue R, Krasny ME, Zazu C (2009) Traditional food crops as a source of community resilience in Zimbabwe. Int J Afr Renaissance Stud 4(1):31–48

Silber T (2013) Kariba REDD+ project monitoring report 2011–2012. South Pole Carbon, Zurich

Simpson MC (2010) Quantification and Magnitude of Losses and Damages Resulting from the Impacts of Climate Change: Modelling the Transformational Impacts and Costs of Sea Level Rise in the Caribbean (Key Points and Summary for Policy Makers Document). New York: United Nations Development Programme (UNDP)

Skutsch MM, Ba L (2010) Crediting carbon in dry forests: the potential for community forest management in West Africa. Forest Policy Econ 12(4):264–270

Speranza CI, Kiteme B, Ambenje P, Wiesmann U, Makali S (2010) Indigenous knowledge related to climate variability and change: insights from droughts in semi-arid areas of former Makueni District, Kenya. Clim Change 100:295–315

SSSI (2018) A rising Africa in a fragile environment the imitative on sustainability, stability and security. Sustainability, Stability and Security Initiative (SSSI). Available online http://www.3S-Initiative.org

Swinkels R, Norman T, Blankespoor B et al (2019) Analysis of spatial patterns of settlement, internal migration, and welfare inequality in Zimbabwe. World Bank Group, Washington, DC

Thinda KT, Ogundeji AA, Belle JA, Ojo TO (2020) Determinants of relevant constraints inhibiting farmers' adoption of climate change adaptation strategies in South Africa. J Asian Afr Stud 56(1):1–18

UNDP (2017) Climate change and human development: towards building a climate resilient nation. UNDP, New York

UNDP (2018) Strengthening biodiversity and ecosystems management and climate-smart landscapes in the mid to lower Zambezi Region of Zimbabwe, New York

UNFCCC (2007) Climate change: impacts, vulnerabilities and adaptation in developing countries. United Nations Framework Convention on Climate Change, Bonn. Available online: https://unfccc.int/resource/docs/publications/impacts.pdf. Downloaded 6 Mar 2020

UN-HABITAT (2017) Trends in urban resilience. United Nations Human Settlements Programme (UN-Habitat), Nairobi. Available online: http://urbanresiliencehub.org/wp-content/uploads/2017/11/Trends_in_Urban_Resilience_2017.pdf. Downloaded 6 Nov 2019

USAID (2018) The intersection of global fragility and climate risks. USAID, Washington, DC

van Niekerk A, Scinocca JF, Shepherd TG (2017) The modulation of stationary waves, and their response to climate change, by parameterized orographic drag. Journal of the Atmospheric Sciences 74(8):2557–2574

Wallner F (2005) Indigenous knowledge and Western science: contradiction or cooperation. Indilinga Afr J Indig Knowl Syst 4(1):46–54

Walsh D (2011) Moving beyond Widdowson and Howard: traditional knowledge as an approach to knowledge. Int J Crit Indig Stud 4(1):2–11

Wang Y, Corson C (2015) The making of a 'charismatic' carbon credit: clean cookstoves and 'uncooperative' women in Western Kenya. 47:2064–2079. https://doi.org/10.1068/a130233p

Widdowson F, Howard A (2008) Disrobing the aboriginal industry: the deception behind indigenous cultural preservation. McGill-Queen's University Press, Montreal

Farmers' Adaptive Capacity to Climate Change in Africa: Small-Scale Farmers in Cameroon

Nyong Princely Awazi, Martin Ngankam Tchamba,
Lucie Felicite Temgoua and Marie-Louise Tientcheu-Avana

Contents

Introduction .
    Background of the Study
Review of Literature
    Perceptions of Climate Change by Small-Scale Farmers in Africa
    Adverse Effects of Climate Change on Africa's Small-Scale Farmers ...
    Drivers of Small-Scale Farmers' Vulnerability to Climate Change in Africa
    Adaptation Options Implemented by Small-Scale Farmers in Africa Confronted with

    Determinants of Small-Scale Farmers' Choice of Adaptive Measures Confronted with

    Barriers to Adaptation for Small-Scale Farmers in Africa Confronted with

    Effectiveness of Small-Scale Farmers' Adaptation Measures in Enhancing Adaptive
    Capacity to Climate Change
Description of Study Area and Methodology
    Description of the Study Area
    Research Methods ....

N. P. Awazi (✉) · M. N. Tchamba · L. F. Temgoua · M.-L. Tientcheu-Avana
Department of Forestry, Faculty of Agronomy and Agricultural Sciences, University of Dschang,
Dschang, Cameroon

    Data Sources and Collection
    Analysis of Data
    Dependent and Independent Variables
Findings
    Variations and Changes in Climate Elements

    Farmer Perceived Factors Influencing Adaptive Capacity to Adverse Climatic Variations
    and Changes
    Farmers' Capacity to Adapt to Climate Change
    Factors Affecting Small-Scale Farmers' Adaptive Capacity to Climate Change

    Variations in Climate Elements
    Adaptive Choices of Small-Scale Farmers Confronted with Climate Change
    Perceived Factors Affecting Farmers' Adaptive Capacity to Climate Change
    Non-Cause-Effect and Cause-Effect Relationship Between Small-Scale Farmers'
    Adaptive Capacity to Climate Change and Independent/Independent Variables
Conclusion and Policy Implications
References

Abstract

Small-scale farmers' limited adaptive capacity confronted with the adversities of climate change is a major call for concern considering that small-scale farms feed over half of the world's population. In this light, small-scale farmers' adaptive choices and adaptive capacity to climate change were assessed. Data were collected from primary and secondary sources using a mixed research approach. Findings revealed that extreme weather events have been recurrent and small-scale farmers perceived access to land, household income, and the planting of trees/shrubs on farms (agroforestry) as the main factors influencing their capacity to adapt to climate change. Agroforestry and monoculture practices were the main adaptive choices of small-scale farmers confronted with climate change. T-test and chi-square test statistics revealed a strong non-cause-effect relationship ($p < 0.001$) between small-scale farmers' capacity to adapt to climate change and different socio-economic, institutional, and environmental variables. Parameter estimates of the binomial logistic regression model indicated the existence of a strong direct cause-effect relationship ($p < 0.05$) between small-scale farmers' capacity to adapt to climate change and access to credit, household income, number of farms, access to information, and access to land, indicating that these variables enhance small-scale farmers' capacity to adapt to climate change. It is recommended that policy makers examine the adaptive choices and determinants of farmers' adaptive capacity unearthed in this chapter when formulating policies geared towards enhancing small-scale farmers' capacity to adapt to climate change.

Keywords

Climate change · Small-scale · Farmers · Adaptive capacity · Africa · Cameroon

Introduction

Background of the Study

The fight against climate change features prominently among the seventeen (17) United Nations Sustainable Development Goals (SDGs) – 2030 Agenda, demonstrating the desire of the global policy making community to tackle climate change, one of the foremost existential threats facing humanity today, head-on (IPCC 2018; Chanana-Nag and Aggarwal 2018; Niles and Salerno 2018). This comes in the wake of unprecedented levels of global warming caused mainly by increasing concentrations of carbon dioxide, methane, nitrous oxides, and other greenhouse gases (GHGs) in the atmosphere (Aggarwal et al. 2015; IPCC 2018). Anthropogenic activities especially excessive fossil fuel combustion, deforestation, and degradation of tropical forests have been singled out as the principal causes of the increasing emissions of greenhouse gases into the atmosphere (Biermann 2007; IPCC 2007; The Royal Society 2010; NAS and RS 2014). With the present climatic variations and changes, humanity has just two choices: adaptation and/or mitigation. With mitigation being a long-term option, adaptation becomes incumbent for different sectors of economic life especially the agricultural sector (Adger et al. 2007; Challinor and Wheeler 2008; Challinor 2009; World Bank 2013; FAO et al. 2018). With the most vulnerable actors in the agricultural sector being small-scale farmers, there is absolute necessity to promote measures that foster adaptation and enhance adaptive capacity to the adversities of climate change.

The FAO (2011) indicated that climate change will seriously threaten the livelihood of small-scale farmers. In 2016, studies demonstrated that small-scale farmers will be adversely affected by changes in climate patterns owing to their limited adaptive capacity (FAO 2016). Small-scale farmers' limited adaptive capacity when confronted with the adversities of climate change is a major call for concern considering that small-scale farmers – who in the majority are found in developing countries – contribute to the nourishment of over half of the world's population (FAO 2016). It is estimated that the developing world has roughly 500 million small-scale farms supporting about two billion people, and these small farms produce about 80% of the food consumed in Asia and sub-Saharan Africa (IFAD 2012). With the number of small-scale farms across the developing world rising (FAO 2010a, b; IPCC 2014; FAO et al. 2018), it becomes necessary to examine the capacity of small-scale farmers to adapt to climate change adversities and to examine the factors influencing the capacity of small-scale farmers to adapt to the negative effects of climate change.

Cameroon like other developing countries is dominated by food-based agricultural systems. These food-based farming systems owned in the majority by small-scale farmers (who constitute over 90% of the farming population) have been adversely affected by climate change (Molua 2006, 2008; Tingem et al. 2009; Azibo and Kimengsi 2015; Awazi 2018). Small-scale farmers' capacity to adapt to climate change could be enhanced if human, material, logistic, and financial resources are placed at their disposal (Molua 2008; Azibo et al. 2016; Innocent et al. 2016). From this perspective, this chapter sought to assess small-scale farmers' adaptive choices and the determinants of small-scale farmers' capacity to adapt to climate change, in the hope that the findings will go a long way to influence policy and alleviate the plight of small-scale farmers.

Review of Literature

Perceptions of Climate Change by Small-Scale Farmers in Africa

Africa's small-scale farmers are increasingly perceptive of climate change, although their perceptions vary on a country-by-country basis as shown by different studies carried out in Africa. In a study carried out by Belaineh et al. (2013) in the Doba District, West Hararghe, Ethiopia, it was found that all male-headed and female-headed households perceived the occurrence of climate change. Boissière et al. (2013) on the contrary, in a study carried out in Indonesia – the tropical forests of Papua – found that the local population's perceptions of adverse climatic variations and changes differed significantly across the studied villages. They concluded that these differences in perception of climate change could be due to the different agro-ecological conditions of the villages. Mtambanengwe et al. (2012) on their part found respondents unanimous that the total quantity of rainfall had declined. The findings of Mtambanengwe et al. (2012) corroborate those of De Wit (2006) and Anderson (2007) who revealed that Southern Africa is becoming increasingly drier, threatening agricultural sustainability, as rainfall distribution within the season fluctuates tremendously.

Maddison (2006), however, found that Zimbabwe's small-scale farmers' perception of climate change varied with respect to the number of years of experience in farming. According to Maddison, small-scale farmers with more than 20 years of experience in farming were more likely to notice significant changes in normal weather patterns compared to their less experienced counterparts. This is corroborated by Mtambanengwe et al. (2012) who also found that 3–4% of small-scale farmers who claimed not to have noticed any shift in climate in the two communities studied in Zimbabwe were young farmers or farmers mostly involved in off-farm activities.

In a study undertaken in South Africa, Benhin (2006) found that about 72% of farmers sampled were of the opinion that climate change has been occurring over the years, with delays in the timing of the rain, a drastic drop in the quantity of rain, and higher temperatures. The farmers' perceptions, however, varied slightly across the nine provinces in which the study was carried out. In the semiarid areas of Tanzania, Mary and Majule (2009) found that 63.8% of farmers sampled in Kamenyanga village and 73.8% of farmers sampled in Kintinku village perceived an increase in temperature. Farmers reported that the months of September to December were becoming extremely hot and the nights were generally becoming very cold. It was also found that most of the farmers sampled perceived a decrease in precipitation and changes in onset of rains as well as an increase in drought frequency in Kamenyanga and Kintinku districts, respectively. Most of the farmers stated that the onset of rainfall has changed because crops were usually planted in the months of October/November but lately crops are being planted in the months of December/January.

In a study undertaken in eleven (11) African countries, Maddison (2006) found that a large majority of farmers believed that precipitation is declining and temperature is on the rise. Majule et al. (2008) also reported similar findings. Tessema et al. (2013) in a study undertaken in the East Hararghe zone of Ethiopia found that farmers' perceptions differ with respect to changes in precipitation and temperature. A large majority (91.2%) of the farmers perceived a rise in temperature, whereas 3.5% and 5.3% of the farmers perceived a decrease in temperature and no change, respectively. Most of the farmers (90.3%) perceived a drop in the quantity of precipitation; meanwhile 2.6% and 6.2% of the farmers perceived an increase in the quantity of precipitation and no change, respectively. Only a small percentage (0.9%) of the farmers indicated that precipitation is variable rather than agreeing either on an increase or decrease in the quantity of rainfall.

In the same line of thought, studies undertaken across different parts of Africa have shown that small-scale farmers perceive climate change through variations in climate elements. Studies undertaken by Ishaya and Abaje (2008) in Kaduna state, Nigeria; Gbetibouo (2009) in the Limpopo Basin of South Africa; Mertz et al. (2009) in the Rural Sahel; Deressa et al. (2011) in the Nile Basin of Ethiopia; Nyanga et al. (2011) in Zambia; Nzeadibe et al. (2012) in the Niger Delta Region of Nigeria; Tambo and Abdoulaye (2012) in Nigeria; Yaro (2013) in Ghana; Juana et al. (2013) in sub-Saharan Africa (synthesis of empirical studies); Temesgen et al. (2014) in Ethiopia; Mulenga and Wineman (2014) in Zambia; and Aggarwal et al. (2015) in the Kullu District of the western Himalayan region all found that small-scale farmers were increasingly perceptive of climate change. Based on the findings of all these studies, a conclusion could be drawn to the effect that small-scale farmers' perceptions of climate change are quasi-unanimous across Africa.

Adverse Effects of Climate Change on Africa's Small-Scale Farmers

Africa's small-scale farmers are increasingly being affected by climate change. Scholarship indicates that climate change has mainly adverse effects on Africa's small-scale farming communities. In a study carried out in Kenya, Herrero et al. (2010) found that climate change adversely affected small-scale farmers through recurrent droughts. Mary and Majule (2009) carried out a study in Tanzania, revealing that the recurrence of extreme climate events (changing rainfall and temperature patterns) led to increased risk of crop failure owing to the washing away of seeds and crops, stunted growth, poor seed germination, and withering of crops. It was equally found that, in the case of livestock, variations in rainfall patterns (decreased rainfall–drought and increased rainfall–floods) led to a decrease in pasture and an increase in parasites and diseases. Similar findings have been reported by other studies carried out in Africa. Mortimore and Adams (2001), for example, found that the timing of the onset of the first rains and other intra-seasonal factors such as the effectiveness of the rains in each precipitation, and the distribution and length of the period of rain during the growing season, seriously affect crop-planting regimes as well as the effectiveness and success of farming. According to the IPCC (2007), changes in rainfall patterns and the quantity of rainfall affect soil moisture and the rate soil erosion, both prerequisites for crop growth and crop yields. All these negatively affect small-scale farmers.

In a study assessing the economic impact of climate change on agriculture in Cameroon, Molua and Lambi (2006) found that as temperature increases, and precipitation decreases, net revenue dropped across all the surveyed farms. The study equally revealed that an increase in temperature by 2.5 °C will lead to a drop in net revenues from agriculture in Cameroon by $0.5 billion. A 5 °C increase in temperature on its part will lead to a drop in net revenues by $1.7 billion. A 7% decrease in precipitation will lead to a drop in net revenues by $1.96 billion, and a 14% decrease in precipitation will lead to a drop in net revenues from crops by $3.8 billion. The study, however, found that increases in precipitation will lead to an increase in net revenues. Based on these findings, small-scale farmers in Cameroon will be adversely affected by climate change through a fall in farm revenue.

On their part, Tabi et al. (2012), in a study carried out in the Volta region of Ghana, found that climate change adversely affects rice farmers. These adverse effects were death of animals, loss of farming capital, heat stress, increase in social vices, shortage of water, slow development, and increased poverty and food insecurity. From these findings and those of other studies aforementioned, it could be said that climate change has mainly adverse or negative effects on small-scale farmers in Africa.

Drivers of Small-Scale Farmers' Vulnerability to Climate Change in Africa

In the face of climate change adversities, small-scale farmers in Africa are the most vulnerable actors involved in the agricultural sector (Rurinda 2014). Small-scale farmers' vulnerability to climate change adversities could be attributed to several factors. In a study carried out to examine the vulnerability of small-scale farming systems of Zimbabwe to climate change, Rurinda (2014) and Rurinda et al. (2014) found that the main causes or sources of vulnerability of small-scale farmers to climate change were lack of knowledge, lack of draught power, increased rainfall variability, lack of seed, lack of fertilizer, and declining soil fertility. Following Rurinda's findings, the single most important source or cause of small-scale farmers' vulnerability to climate change was increasing variability in rainfall.

In a study assessing rice farming in the Volta region of Ghana, Tabi et al. (2012) showed that the main sources or causes of rice farmers' vulnerability to climate change were low price of rice in the local market, difficult land tenure system, limited or no access to credit facilities, few farmers engaged in off-farm activities, poor soils, and lack of insurance in times of crop failure.

The CDCCP (2009), in a study undertaken in the Chiredzi district of Zimbabwe, found that the main causes of small-scale farmers' vulnerability to climate change were as follows: poor farming practices, high frequency of drought, inherent dryness, limited use of climate early warning systems, over-dependence on mono-cropping especially maize, high incidence of poverty, population pressure, skewed ownership and access to drylands' livelihood assets such as livestock and wildlife, lack of drought preparedness plans, limited alternative livelihood options outside agriculture, and low access to technology (irrigation, seed), markets, institutions, and infrastructure (poor roads, bridges, modern energy, dams and water conveyance).

These findings therefore demonstrate that small-scale farmers in Africa are highly vulnerable to the adverse effects of climate change.

Adaptation Options Implemented by Small-Scale Farmers in Africa Confronted with Climate Change

In Africa, small-scale farmers have adopted different adaptive options in order to improve their adaptive capacity confronted with climate change. Tabi et al. (2012) while assessing rice farming in the Volta region of Ghana found that rain-fed lowland rice farmers practiced different adaptive choices among which were the application of fertilizers, water management control practices, alternation of planting dates, herbicide use, and the use of high-yielding and disease-resistant varieties. On their part, Kuwornu et al. (2013) in a study carried out in northern Ghana found that small-scale farmers adopted both indigenous and introduced (modern) adaptive options to improve their adaptive capacity to climate change.

Molua and Lambi (2006), in a study undertaken in Cameroon, found that the main indigenous adaptation strategies implemented by small-scale farmers in the face of climate change were changing timing of farming operations, increasing planting space, undertaking traditional and religious ceremonies, change of crops, varying area cultivated, and cultivation of short season local varieties. The FAO (2006) found that the major indigenous adaptation strategies practiced by small-scale farmers were reducing food intake, change of crops, reducing personal expenditures, mortgaging land, homestead gardening, disposing of productive harvests, and re-sowing or re-planting.

Different authors have carried out studies across Africa with varying findings as far as indigenous adaptive choices implemented by small-scale farmers confronted with climate change adversities are concerned. For example, studies carried out by Hassan and Nhemachena (2008), the FAO (2008, 2009b), Gbetibouo (2009), and Deressa et al. (2010) showed that diversification of crops is a major indigenous strategy practiced by small-scale farmers confronted with climate change adversities. Studies carried out by Easterling et al. (2007), Boko et al. (2007), Gbetibouo (2009), the FAO (2009a, 2010c), and Deressa et al. (2010) showed that the integration of livestock to crop production is a key indigenous strategy practiced by small-scale farmers confronted with climate change. Studies undertaken by Molua and Lambi (2006), Easterling et al. (2007), Boko et al. (2007), Hassan and

Nhemachena (2008), Gbetibouo (2009), and the FAO (2009b) revealed that changing the timing of farm operations is one of the most important indigenous strategies adopted by small-scale farmers in the face of climate change. The FAO (2006, 2009b), Molua and Lambi (2006), and Gbetibouo (2009) found that changing of crops was a major adaptation strategy used by small-scale farmers to adapt to climate change. The FAO (2006) and Altieri and Koohafkan (2008) found that home gardening was a major indigenous strategy practiced by small-scale farmers confronted with climate change adversities.

The FAO (2010a), Thorlakson (2011), Rao et al. (2011), Mbow et al. (2013), Bishaw et al. (2013), Mbow et al. (2014), Kabir et al. (2015) and Awazi and Tchamba (2019), found that agroforestry practices like scattered trees on croplands, improved fallows, home gardens, and cocoa, coffee, and banana agroforests were sustainable and climate-smart adaptive choices practiced by small-scale farmers across Africa in the face of climate change.

From the foregoing, small-scale farmers are adopting both indigenous and introduced adaptive measures to adapt to the adverse effects of climate change across Africa. However, very little has been done to assess the adaptive capacity of small-scale farmers in the face of climate change.

Determinants of Small-Scale Farmers' Choice of Adaptive Measures Confronted with Climate Change

Small-scale farmers' choice of adaptive measures confronted with climate change was influenced by several factors. Tabi et al. (2012), in a study carried out to assess the different adaptive choices of small-scale rice farmers in Ghana, found that the main variables influencing the different adaptive options of small-scale farmers were distance to farm and market, labor, advice from extension agents, gender, length of stay in rice farm, age, farm size, number of farms, credits, household size, and education. Deressa et al. (2008) and Atinkut and Mebrat (2016) on their part found that different infrastructural and institutional factors as well as household and farm characteristics influenced the adaptive choices of small-scale farmers confronted with adverse climatic changes. Through marginal analysis, Deressa et al. (2008), in a study carried out in the Nile Basin of Ethiopia, found that institutional factors (availability of information), social capital, household variables, agro-ecological features, and wealth attributes influenced small-scale farmers' adaptive choices confronted with climate change in the Nile Basin of Ethiopia.

Studies carried out by Maddison 2006 and Nhemachena and Hassan (2007) showed that the most common household attributes influencing small-scale farmers' adaptive capacity to climate change adversities were wealth, marital status, farming experience, age, education, and gender of the head of household; common farm attributes influencing small-scale farmers' adaptive choices to climate change included fertility, slope, and farm size; common institutional factors affecting adaptive choices of small-scale farmers to adverse climatic changes included credit accessibility and access to extension services; and the common infrastructural factor influencing small-scale farmers' adaptive capacity to climate change was distance to input and output markets. Across Africa, different studies have been carried out assessing the impact of climate change and factors affecting small-scale farmers' adaptive choices in crop, livestock, and mixed crop-livestock production systems confronted with climate change adversities (Maddison 2006; Hassan and Nhemachena 2008; Kurukulasuriya and Mendelsohn 2007). However, Zivanomoyo and Mukarati (2012) assessed the factors affecting small-scale farmers' choice of crop varieties confronted with climate change adversities. The study sought to examine how farmers' choice of different crop varieties contributed to improve their adaptive capacity to climate change. Findings revealed that the use of more disease-resistant and hybrid varieties contributed in a major way towards enhancing the adaptive capacity of small-scale farmers to climate change.

From the foregoing, the factors influencing the adaptive choices of small-scale farmers confronted with climate change adversities could be broadly classified into institutional, environmental, and socio-economic factors. Although the factors influencing the adaptive choices of small-scale farmers confronted with climate change have been examined by different studies across Africa, little has been done to assess the small-scale farmers' adaptive capacity to climate change.

Barriers to Adaptation for Small-Scale Farmers in Africa Confronted with Climate Change

In Africa, small-scale farmers have increasingly faced difficulties adapting to the adverse effects of climate change because of different factors. Tabi et al. (2012), in a study carried out on small-scale rice farmers in the Upper Volta region of Ghana, found that the main barriers to small-scale farmers' adoption of different adaptive options confronted with climate change were lack of equipment for quick and appropriate land preparation, lack of farm inputs, inadequate or no irrigation facilities, inadequate or no weather forecast, and limited access to credits. Deressa et al. (2008) in a study carried out in the Nile Basin of Ethiopia found that five major constraints affected small-scale farmers' adaptation choices to climate change. These barriers were shortage of land, shortage of labor, poor potential for irrigation, lack of money, and lack of information. The study however found that most of the constraints to small-scale farmers' adaptation to climate change could be largely attributed to poverty. Deressa et al. (2009, 2011), in studies undertaken in the Nile Basin of Ethiopia, equally demonstrated that poverty is a major barrier to small-scale farmers' adaptation to climate change, because lack of money makes it difficult for small-scale farmers to get the required resources and technologies that ease adaptation to climate change.

In a study carried out in the coastal regions of Bangladesh, Kabir et al. (2015) found that the main constraints to climate change adaptation for small-scale farmers were lack of information, lack of credit, unpredicted weather, shortage of land, shortage of farm inputs, and lack of water. Tessema et al. (2013), in a study carried out in the Eastern Hararghe Zone of Ethiopia, discovered that the major constraints to climate change adaptation for small-scale farmers were shortage of land, lack of seed, shortage of labor, limited market access, lack of money, lack of water, lack of fertilizer, lack of oxen, insecure land tenure, and lack of information. In different parts of Ethiopia, studies have shown that small-scale farmers face several difficulties in their drive to adapt to climate change (Maddison 2007; Deressa et al. 2009, 2011; Bryan et al. 2009; Mersha and Laerhoven 2016).

The aforementioned studies indicate that, small-scale farmers' inability to adapt to climate change is largely due to different barriers. However, limited work has been done to examine the adaptive capacity of small-scale farmers confronted with climate change and the barriers to small-scale farmers' capacity to adapt to climate change.

Effectiveness of Small-Scale Farmers' Adaptation Measures in Enhancing Adaptive Capacity to Climate Change

In Africa, the effectiveness of small-scale farmers' adaptive choices confronted with climate change varies tremendously. Kuwornu et al. (2013) carried out a study to assess the adaptive options of small-scale farmers confronted with climate change adversities and the effectiveness of these adaptive options. They found that among the different indigenous strategies used by small-scale farmers to adapt to the adversities of climate change, the strategies comprising of timing of rainfall and early or late planting were ranked by small-scale farmers in northern Ghana as the most effective strategy used in adapting to adverse climate change while soil-related strategies were ranked as the least effective indigenous strategy used by small-scale farmers. Kuwornu et al. (2013) equally found that among the introduced adaptation strategies (adaptation strategies introduced by research), soil and plant health strategies were ranked by small-scale farmers as the most effective introduced strategy enhancing adaptation to climate change, while non-adoption of any of the introduced strategies was quasi-unanimously ranked by small-scale farmers as the least effective way of adapting to the adverse effects of changing climatic conditions. Hadgu et al. (2015) in a study undertaken in the Tigray region of northern Ethiopia found that changing of crop variety/type was ranked by small-scale farmers as the most effective adaptation strategy to climate change while the "No" adaptation strategy was the least effective way to adapt to the adversities of changing climatic conditions.

The review of previous literature enabled the authors of this chapter to understand what had been done on the African continent in general and Cameroon in particular as far as small-scale farmers' adaptation and adaptive capacity to climate change were concerned. It equally afforded the authors the opportunity to identify some independent variables used in the chapter. However, it was found that while many authors had undertaken studies which revealed that small-scale farmers adopt different adaptive choices in the face of climate change, little had been done to examine small-scale farmers' adaptive capacity to climate change. This work was therefore initiated in a bid to fill the knowledge gap.

Description of Study Area and Methodology

Description of the Study Area

The study was carried out in the north-west region of Cameroon. The north-west region of Cameroon lies between longitude 9°30′E to 11°15′E and latitude 5°4′N to 7°15′N. The north-west region of Cameroon covers a total surface area of about 17,812 km^2 and hosts a population of over 1,840,500 inhabitants, which gives a population density of roughly 103 inhabitants/km^2 making it one of the most densely populated regions in Cameroon. The climate is tropical, and the vegetation is mostly made up of savannah grassland, interspersed with some forest patches. The topography is rolling and characterized by mountains like Mount Oku and plains like the Ndop and Mbaw plains.

Research Methods

Study Site Selection and Sampling Methods

The multiphase sampling procedure was employed. At the first phase, the area of study (the north-west region of Cameroon) was selected purposively owing to the presence of mainly small-scale farmers and the high levels of vulnerability of these small-scale farmers to climate change. At the second phase, ten villages were randomly selected from the different sub-districts found in the north-west region of Cameroon, taking into cognizance the agro-ecological, socio-economic, and environmental attributes of the different sub-districts. This was done with the help of agricultural extension agents working in the area. The third phase involved focus group discussions with small-scale farmers and key informant interviews with resource persons. This was done in order to get general information on the adaptive capacity of small-scale farmers and to triangulate this information with that gotten from small-scale farmers during household surveys. In the fourth and last phase, household surveys were conducted in the ten villages using the simple random sampling approach. With the use of semi-structured questionnaires, a total of 350 small-scale farmer household heads were sampled in the ten villages.

Data Sources and Collection

Both secondary and primary data were collected. Secondary data were collected primarily through the review of relevant literature from previous scientific studies as well as climate data from meteorological stations in the study area. Primary data were collected through a survey of 350 small-scale farmer household heads, complemented with focus group discussions, key informant interviews with resource persons, and overt observations.

Through the use of five-point Likert scale-style questions during household surveys, farmers were asked to rank their adaptive capacity to climate change based on their livelihood capital assets. These livelihood capital assets were natural, human, social, financial, and physical. These different capital assets constituted the independent variables of the study. It was on the basis of these capital assets that farmers ranked their adaptive capacity to climate change to be high, low, or no adaptive capacity.

Analysis of Data

Primary data collected on the field was coded and imputed into Microsoft Excel 2007 and SPSS 20.0 statistical packages for descriptive and inferential statistical analysis. Descriptive statistics computed were charts and percentage indices, while inferential statistics computed were t-test statistic, chi-square test, and logistic regression.

The independent samples t-test and chi-square (X^2) test statistics were used to identify the non-cause-effect relationship between small-scale farmers' capacity to adapt to climate change and independent variables.

The binary logistic (BNL) regression model on its part was used to examine the cause-effect relationship between small-scale farmers' capacity to adapt to climate change and independent variables. The binary logistic regression model predicts the log ODDS of having made one decision or the other. This model permits the analysis of decisions across two categories (Di Falcao et al. 2011; Awazi and Tchamba 2018).

Dependent and Independent Variables

Both dependent and independent variables were used. The dependent variable was adaptive capacity (binary, i.e., adaptive/not adaptive), while the independent or independent variables (different capital assets) were age of household head, household size, number of farms, income of household, educational level, gender of household head, practice of agroforestry, vulnerability to climate change, information accessibility, credit accessibility, land accessibility, and access to extension services. Because the dependent and independent variables were mainly qualitative in nature, the statistical analyses were done using non-parametric tests and the discrete regression model (binomial logistic regression).

Findings

Variations and Changes in Climate Elements

The analysis of over five decades of climate data revealed significant variations in climate parameters (Figs. 1, 2, and 3). In the past 58 years (1961–2018), temperature fluctuations were high, and most of the years experienced above mean temperature, implying that temperature is becoming higher than usual. Meanwhile the total

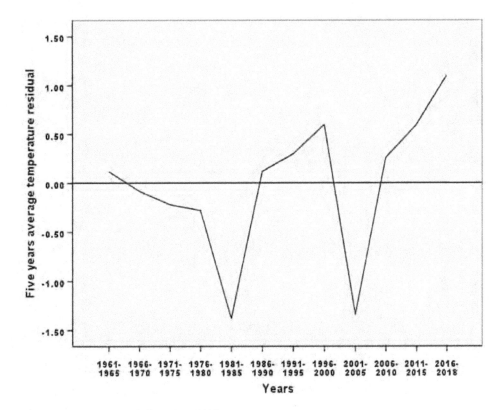

Fig. 1 Temperature variation 1961–2018

quantity of rainfall and number of rainy days equally experienced marked levels of fluctuation, with most of the years experiencing a decrease in amount of rainfall and fewer rainy days. This indicates that the amount of rainfall has been scanty while the number of rainy days has been erratic. These high levels of fluctuation in climate parameters within the past 58 years could therefore be seen as an indicator of climatic variations and changes.

In the face of climatic variations and changes, the relationship between the different climate parameters varied. Scatter plots indicated the existence of an insignificant negative correlation between rainfall and temperature, and rainy days and temperature. Meanwhile a relatively strong positive correlation was found to exist between rainfall and rainy days.

Adaptive Choices of Small-Scale Farmers Confronted with Climate Change Adversities

An analysis of small-scale farmers' adaptive choices confronted with climate change showed that a majority of the small-scale farmers (74%) were practicing agroforestry on their farm plots (Table 1). Among the agroforestry practices most patronized by

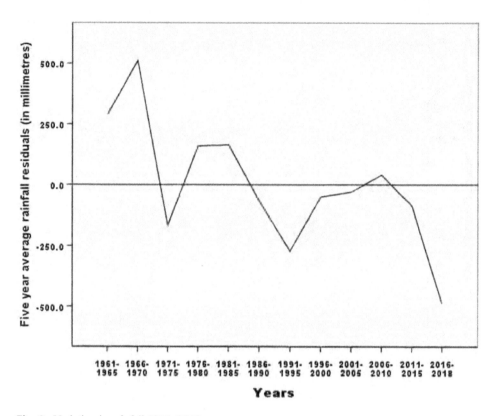

Fig. 2 Variation in rainfall 1961–2018

small-scale farmers confronted with adverse climatic variations and changes were home gardens with livestock (13%), home garden (11%), trees on croplands (11%), live fences/hedges (11%), and coffee-based agroforestry (9%) (Table 1).

Equally, some small-scale farmers confronted with adverse climatic variations and changes practiced monoculture (Table 1). The most common monoculture and mono-livestock practices of small-scale farmers confronted with adverse climatic variations and changes were market gardening monoculture (8%), cash crop mono-culture (7%), and food crops monoculture (9%).

Farmer Perceived Factors Influencing Adaptive Capacity to Adverse Climatic Variations and Changes

Assessing small-scale farmers' adaptive capacity to adverse climatic variations and changes (Fig. 4), it was found that all the small-scale farmers perceived land accessibility (100%) and income of household (100%) as being the main factors influencing adaptive capacity to climate change.

Agroforestry (82%), accessibility to markets (77%), credit accessibility (72%), information accessibility (65%), and access to extension services (55%) were

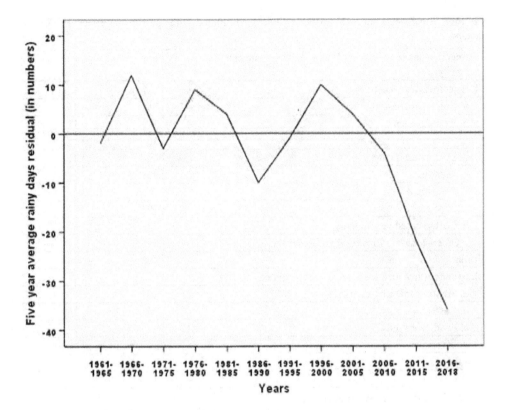

Fig. 3 Variation in rainy days 1961–2018

equally perceived by small-scale farmers as being among the key factors affecting adaptive capacity to climate change. Other least perceived factors influencing adaptive capacity to climate change were irrigation (31%) and others (14%) like road network and topography. However, it is worth mentioning that the main factors influencing small-scale farmers' adaptive capacity to climate change were land accessibility, income of household, agroforestry, accessibility to markets, credit accessibility, and information accessibility (Fig. 4).

Farmers' Capacity to Adapt to Climate Change

Concerning the adaptive capacity of small-scale farmers to climate change (Fig. 5), most small-scale farmers perceived that on the basis of their livelihood capital assets, they were not adaptive (58%).

Meanwhile 14%, 20%, and 4% of small-scale farmers perceived that, on the basis of their livelihood capital assets, they were adaptive, less adaptive, and much less adaptive, respectively, to climate change. Only 4% of small-scale farmers perceived that, on the basis of their livelihood capital assets, they were highly adaptive to climate change. From these perceptions, it was noticed that most small-scale farmers had a limited capacity to adapt to climate change (Fig. 5).

Table 1 Small-scale farmers' adaptive choices confronted with the adverse effects of climate change

Farmers' adaptive choice confronted with climate change adversity	Frequency (n)	Percent (%)
1. Agroforestry practices		
a. Home garden with livestock	35	13
b. Home garden	30	11
c. Trees on croplands	30	11
d. Live fences/hedges	30	11
e. Taungya	20	7
f. Trees on grazing lands	15	6
g. Improved fallows	10	4
h. Coffee-based agroforestry	25	9
i. Others (entomoforestry, aquaforestry)	5	2
Total	**200**	**74**
2. Monoculture and mono-livestock practices		
a. Market gardening crops only	20	8
b. Cash crops only	20	7
c. Food crops only	25	9
d. Livestock only	5	2
Total	**70**	**26**
N	**270**	**100**

Source: Adapted from Awazi et al. 2020

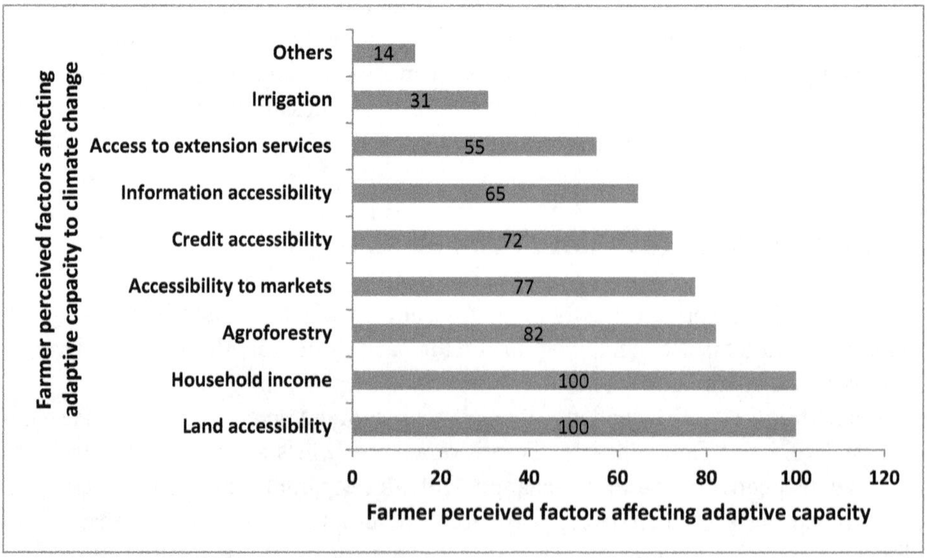

Fig. 4 Factors influencing adaptive capacity to climate change perceived by small-scale farmers

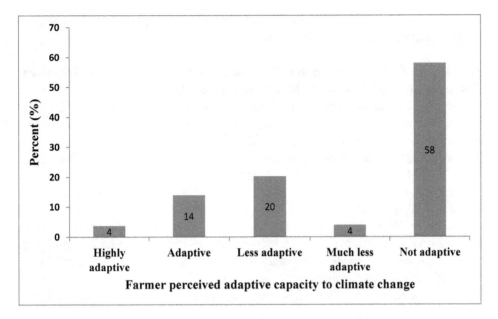

Fig. 5 Adaptive capacity to climate change perceived by small-scale farmers

Table 2 Non-cause-effect relationship between small-scale farmers' adaptive capacity to climate change and four continuous independent variables

Independent variable	T-test for equality of means			
	t	df	p-level	Mean diff.
Number of farms	−10.776	170.493	0.000[***]	−2.940
Household size	−7.552	195.262	0.000[***]	−1.590
Age of household head	−8.224	209.441	0.000[***]	−5.192
Income of household (in FCFA)	−9.062	179.442	0.000[***]	−179415.9

[***]Significant at 1% probability level

Factors Affecting Small-Scale Farmers' Adaptive Capacity to Climate Change

Non-Cause-Effect Relationship Between Small-Scale Farmers' Adaptive Capacity and Continuous Independent Variables

T-test statistics showed that there was a significant non-cause-effect relationship between small-scale farmers' adaptive capacity and four continuous independent variables (Table 2).

The continuous independent variables (number of farms ($t = 10.776, p < 0.001$), size of household ($t = 7.552, p < 0.001$), age of household head ($t = 8.224, p < 0.001$), and income of household ($t = 9.062, p < 0.001$)) all had a significant non-cause-effect relationship with small-scale farmers' adaptive capacity to climate change. This demonstrates that the number of farms owned, size of household, age of household head, and income of household play an important role in influencing the adaptive capacity of small-scale farmers confronted with climate change.

Non-Cause-Effect Relationship Between Small-Scale Farmers' Adaptive Capacity and Qualitative Independent Variables

Chi-square test statistics showed that there was a significant non-cause-effect relationship between small-scale farmers' adaptive capacity to climate change and seven (07) qualitative independent variables (Table 3).

The qualitative independent variables (level of education of household head ($X^2 = 123.10$, $p < 0.001$), gender of household head ($X^2 = 24.95$, $p < 0.001$), practice of agroforestry ($X^2 = 64.50$, $p < 0.001$), information accessibility ($X^2 = 44.70$, $p < 0.001$), credit accessibility ($X^2 = 90.88$, $p < 0.001$), land accessibility ($X^2 = 52.50$, $p < 0.001$), and access to agricultural extension services ($X^2 = 21.54$, $p < 0.001$)) all had a significant non-cause-effect relationship with small-scale farmers' adaptive capacity to climate change. This confirms that the level of education of household head, gender of household head, practice of agroforestry, access to information, access to credit, access to land, and access to

Table 3 Non-cause-effect relationship between small-scale farmers' adaptive capacity and qualitative independent variables

Qualitative independent variable	Description	Frequency (n)		Percentage (%)		Chi-square	L.R.	p-level
		A.	N. A.	A.	N.A.			
Educational level of household head	No formal edu.	11	21	3.14	6	123.10	141.69	0.000***
	Primary	62	180	17.71	51.43			
	Secondary	10	0	2.86	0			
	High school	34	1	9.71	2.86			
	Tertiary	30	1	8.57	2.86			
Gender of household head	Male	104	89	29.71	25.43	24.95	25.47	0.000***
	Female	43	114	12.28	32.57			
Practice agroforestry	Yes	147	132	42	37.71	64.50	90.23	0.000***
	No	0	71	0	20.28			
Information accessibility	Yes	42	7	12	2	44.70	46.69	0.000***
	No	105	196	30	56			
Credit accessibility	Yes	64	5	18.28	1.43	90.88	99.25	0.000***
	No	83	198	23.71	56.57			
Land accessibility	Yes	51	10	14.57	2.86	52.50	54.33	0.000***
	No	96	193	27.43	55.14			
Access to extension	Yes	45	22	12.86	6.29	21.54	21.41	0.000***
	No	102	181	29.14	51.71			

***Significant at 1% probability level; A. = adaptive; N.A. = not adaptive; L.R. = likelihood ratio

agricultural extension services influence small-scale farmers' adaptive capacity to climate change.

Binary Logistic Regression Model Predicting Small-Scale Farmers' Adaptive Capacity to Climatic Change from Independent Variables

The parameter estimates of the binary logistic regression model revealed that five main independent variables played a statistically significant role in influencing small-scale farmers' adaptive capacity to climate changes (Table 4).

From the parameter estimates of the binary logistic regression model, the number of farms ($\beta = 0.271, p < 0.05$), information accessibility ($\beta = 0.937, p < 0.1$), credit accessibility ($\beta = 1.596, p < 0.05$), income of household ($\beta = 1.821, p < 0.01$), and land accessibility ($\beta = 1.029, p < 0.05$) all had a significant direct cause-effect relationship with small-scale farmers' adaptive capacity to climate change. This implies that as the number of farms, information accessibility, credit accessibility, household income, and land accessibility increase, small-scale farmers' adaptive capacity to climate change also increases.

It is important to note that the parameter estimates of this model were valid looking at the likelihood ratio X^2, the number of cases correctly classified, and the Nagelkerke R^2. The likelihood ratio X^2 ($5, n = 350 = 145.835, p < 0.01$) indicated that the model was statistically significant and had a strong explanatory power. The model correctly classified up to 80% of the factors influencing small-scale farmers' adaptive capacity to climate change. Looking at the Nagelkerke R^2 (Pseudo R^2) of the model which stood at 0.648, it revealed that up to 64.8% of the changes in small-scale farmers' adaptive capacity to climate change could be explained by changes in the continuous and qualitative independent variables of the model. Hence, from the values of the likelihood ratio X^2, the number of cases correctly classified, and the Nagelkerke R^2, it could be said that the predictions of the model were very much

Table 4 Logistic regression showing influence of independent variables on the adaptive capacity of small-scale farmers to climate change

Independent variables	Coefficients (β)	p-level	Std error	Wald	df	Odds ratio (Exp β)
Constant	−1.961***	0.000	0.294	44.426	1	0.141
Number of farms	0.271**	0.003	0.092	8.690	1	1.311
Income of household	1.821***	0.002	0.614	9.064	1	5.134
Information accessibility	0.937*	0.087	0.548	2.929	1	2.553
Credit accessibility	1.596**	0.006	0.582	7.526	1	4.931
Land accessibility	1.029**	0.027	0.465	4.891	1	2.798
Log likelihood	330.37					
Likelihood ratio X^2	145.84***	0.000				
Nagelkerke R^2	0.648					
Number of cases correctly classified	80%					

*, **, ***Significant at 10%, 5%, and 1% probability levels, respectively

valid as far as determining the factors influencing small-scale farmers' adaptive capacity to climate change was concerned.

Discussion

Variations in Climate Elements

Extreme levels of variation in climate parameters (rainfall, temperature, and rainy days) have been recurrent in the north-west region of Cameroon in the past five decades which attests to climate variations and changes. Although other studies have shown the occurrence of climate variability in the north-west region of Cameroon (Innocent et al. 2016; Awazi 2018; Awazi and Tchamba 2018; Awazi et al. 2019), few studies in Cameroon have examined climatic variations using over five decades of climate data.

The chapter equally assessed the non-cause-effect and cause-effect relationship existing between climate parameters (temperature, rainy days, and rainfall) in the face of adverse climatic variations and changes. It was found that there is a very limited inverse relationship between rainfall and temperature as well as rainy days and temperature. Meanwhile a relatively strong direct relationship exists between rainfall and rainy days. This indicated that an interdependent relationship exists between rainfall and rainy days in the face of climate change. Studies carried out in other parts of the world by Chen and Wang (1995), Buishand and Brandsma (1999), Seleshi and Zanke (2004), Cong and Brady (2012), Berg et al. (2013), Olsson et al. (2015), Nkuna and Odiyo (2016), and Weng et al. (2017) have equally proven the existence of an interdependent relationship between climate parameters, although not in the context of climate change. By examining the relationship between climate parameters in the face of climate change, this chapter has filled a major knowledge gap.

Adaptive Choices of Small-Scale Farmers Confronted with Climate Change

It was found that most small-scale farmers practice agroforestry to enhance their adaptive capacity to climate change. Agroforestry practices therefore constitute a major adaptive choice for small-scale farmers. Most studies carried out in Africa (Easterling et al. 2007; Boko et al. 2007; Hassan and Nhemachena 2008; Gbetibouo 2009; FAO 2009b, 2010; Deressa et al. 2010) have merely shown that small-scale farmers adopt indigenous and non-indigenous adaptation strategies to combat the adversities of climate change. This chapter revealed that most small-scale farmers adopt agro-ecological farming practices like agroforestry to mitigate the adverse effects of climate change, thereby filling the knowledge gap.

Perceived Factors Affecting Farmers' Adaptive Capacity to Climate Change

Small-scale farmers perceived several factors influencing their adaptive capacity. Among these factors, land accessibility, income of household, and the practice of agroforestry were perceived by small-scale farmers as the most important factors affecting their adaptive capacity to climate change. As reported by other scientific studies, small-scale farmers usually perceive a combination of factors influencing their adaptation to climate variations and changes. Most studies have assessed the different adaptation choices practiced by small-scale farmers to enhance adaptive capacity to climate change (McCarthy et al. 2004; Gbetibouo and Ringler 2009; Folke et al. 2010; World Bank 2013), with very limited literature dwelling on the adaptive capacity of small-scale farmers confronted with climate change. By examining small-scale farmers' perceptions of their adaptive capacity to climate change, this chapter has filled the knowledge gap. Although some studies (Gordon 2009; Gbetibouo 2009; Thorlakson 2011) have used conceptual and theoretical approaches to assess adaptive capacity to climate change, this chapter by applying the inferential statistical approach to examine small-scale farmers' adaptive capacity to climate change adversities has filled a major knowledge gap.

Non-Cause-Effect and Cause-Effect Relationship Between Small-Scale Farmers' Adaptive Capacity to Climate Change and Independent/ Independent Variables

A non-cause-effect relationship was found to exist between small-scale farmers' adaptive capacity and independent variables (institutional, environmental, and socio-economic variables) like number of farms, household size, age of household head, income of household, level of education, gender, practice of agroforestry, information accessibility, credit accessibility, land accessibility, and access to extension services. Most studies undertaken across Africa and different parts of the world (McCarthy et al. 2004; Gbetibouo and Ringler 2009; Gordon 2009; Gbetibouo 2009; Folke et al. 2010; Thorlakson 2011; World Bank 2013; Awazi 2018; Awazi et al. 2019) mainly examined the non-cause-effect relationship existing between independent variables and small-scale farmers' adaptation choices to climate change. By unraveling the existence of a non-cause-effect relationship between adaptive capacity and different independent variables (institutional, environmental, and socio-economic variables), this chapter has therefore filled a knowledge gap.

A direct cause-effect relationship was found to exist between small-scale farmers' adaptive capacity and five main independent variables (credit accessibility, information accessibility, income of household, number of farms, and land accessibility). These five independent variables could therefore be considered as very important in enhancing small-scale farmers' adaptive capacity to climate change.

Thus, small-scale farmers with many farms are more adaptive to climate change which could be attributed to more yields obtained from these many farms which are consumed by the household and the excess sold to buy farm inputs. It could equally be that these farmers have more access to social and financial resources and/or better education which allows them to control more land and therefore enhanced adaptive capacity.

In the same light, small-scale farmers with better information accessibility are more adaptive to climate change than their counterparts with limited or no access which could be attributed to the fact that small-scale farmers with easy access to information are able to make plans into the future which helps them to adopt best practices.

Equally, small-scale farmers with more access to credit are more adaptive to climate change adversities than their fellow farmers with limited or no access to credit. This could be due to the fact that small-scale farmers with easy access to credit facilities are able to buy better farm inputs and can easily switch to best practices which act as a buffer to the adverse effects of climate change. Meanwhile small-scale farmers with little or no access to credit facilities are unable to buy good farm inputs and cannot switch to best practices on time which renders them weak and vulnerable in the face of climatic extremes.

Similarly, small-scale farmers with more access to land are better adaptive to climate change than their counterparts with limited or no land which can be attributed to the fact that land is an indispensable asset to any farmer, for it is the most important fixed asset, and without it, no farming activity can take place.

Some authors like Mccarthy et al. (2004), Gbetibouo (2009), Thorlakson (2011), and Awazi et al. (2019) have found the existence of cause-effect relationship between small-scale farmers' adaptation choices to climate change and different independent variables. Through the use of inferential statistics to examine the cause-effect relationship between adaptive capacity and different independent variables (institutional, socio-economic, and environmental factors), this chapter fills a major knowledge gap.

Conclusion and Policy Implications

Based on the findings of this chapter, it is clearly noticed that institutional, socio-economic, and environmental factors are key determinants of small-scale farmers' adaptive capacity to climate change. The existence of a statistically significant direct cause-effect relationship between small-scale farmers' adaptive capacity and independent variables such as credit accessibility, information accessibility, land accessibility, income of household, and number of farms is testament to the vital role these livelihood capital assets play towards enhancing small-scale farmers' adaptive capacity to climate change. Thus, it is recommended that policy makers seeking to alleviate the plight of vulnerable small-scale farmers take these determinants of small-scale farmers' adaptive capacity into consideration when formulating policies geared towards enhancing small-scale farmers' adaptive capacity to climate change.

References

Adger WN, Agrawala S, Mirza MMQ, Conde C, O'Brien K, Puhlin J, Pulwarty R, Smit B, Takahashi K (2007) Assessment of adaptation practices, options, constraints and capacity. In climate change 2007: impacts, adaptation and vulnerability. Contribution of Working Group II to the fourth assessment of the IPCC, Palutikof JP, van der Linden PJ, Hansen CE (eds). Cambridge University Press, Cambridge. http://www.ipcc-wg2.gov/AR4/website/17.pdf

Aggarwal VRK, Mahajan PK, Negi YS, Bhardwaj SK (2015) Trend analysis of weather parameters and people perception in Kullu District of Western Himalayan region. Environ Ecol Res 3 (1):24–33. https://doi.org/10.13189/eer.2015.030104

Altieri MA, Koohafkan P (2008) Enduring farms: climate change, small-scales and traditional farming communities. Environment and development series 6. Third world network. Penang, Malaysia. http://www.fao.org/nr/water/docs/enduring_farms.pdf

Anderson J (2007) How much did property rights matter? Understanding food insecurity in Zimbabwe: a critique of Richardson. Afr Aff 106(425):681–690. https://doi.org/10.1093/afraf/adm064.pdf

Atinkut B, Mebrat A (2016) Determinants of farmers' choice of adaptation to climate variability in Dera Woreda, South Gondar Zone, Ethiopia. Environ Syst Res 5:6. http://link.springer.com/content/pdf/10.1186%2Fs40068-015-0046-x.pdf

Awazi NP (2018) Adaptation options enhancing farmers' resilience to climate change. LAP LAMBERT Academic, 132 p. ISBN-10: 3330027940; ISBN-13: 978-3330027947

Awazi NP, Tchamba NM (2018) Determinants of small-scale farmers' adaptation decision to climate variability and change in the north-west region of Cameroon. Afr J Agric Res 13:534–543

Awazi NP, Tchamba NM (2019) Enhancing agricultural sustainability and productivity under changing climate conditions through improved agroforestry practices in small-scale farming systems in sub-Saharan Africa. Afr J Agric Res 14(7):379–388

Awazi NP, Tchamba NM, Tabi FO (2019) An assessment of adaptation options enhancing small-scale farmers' resilience to climate variability and change: case of Mbengwi Central sub-Division, the north-west region of Cameroon. Afr J Agric Res 14(6):321–334

Awazi NP, Tchamba NM, Temgoua LF (2020) Enhancement of resilience to climate variability and change through agroforestry practices in smallholder farming systems in Cameroon. Agrofor Syst 94:687–705. https://doi.org/10.1007/s10457-019-00435-y

Azibo BR, Kimengsi JN (2015) Building an indigenous agro-pastoral adaptation framework to climate change in sub-Saharan Africa: experiences from the the north-west region of Cameroon. Procedia Environ Sci 29:126–127

Azibo BR, Kimengsi JN, Buchenrieder G (2016) Understanding and building on indigenous agro-pastoral adaptation strategies for climate change in sub-Saharan Africa: experiences from rural Cameroon. J Adv Agric 6(1):833–840. https://doi.org/10.24297/jaa.v6i1.5391

Belaineh L, Ayele Y, Bewket W (2013) Smallholder farmers' perceptions and adaptation to climate variability and change in Doba District, West Hararghe, Ethiopia. Asian J Empir Res 3(3):251–265. http://www.aessweb.com/download.php?id=1838.pdf

Benhin J (2006) Climate change and South African agriculture: impacts and adaptation options. CEEPA discussion paper no 21, July 2006. https://www.weadapt.org/sites/weadapt.org/files/legacy-new/knowledge-base/files/5370f181a5657504721bd5c21csouth-african-agriculture.pdf

Berg P, Moseley C, Haerter JO (2013) Strong increase in convective precipitation in response to higher temperatures. Nat Geosci 6:181–185

Biermann F (2007) Earth system governance as a crosscutting theme of global change research. Global Environ Chang 17(2007):326–337. http://www.glogov.org/images/doc/Biermann%202007%20GEC%20Earth%20System%20Governance.pdf

Bishaw B, Neufeldt H, Mowo J, Abdelkadir A, Muriuki J, Dalle G, Assefa T, Guillozet K, Kassa H, Dawson IK, Luedeling E, Mbow C (2013) Farmers' strategies for adapting to and mitigating climate variability and change through agroforestry in Ethiopia and Kenya. Davis CM, Bernart B, Dmitriev A (eds). Forestry Communications Group, Oregon State University, Corvallis. http://international.oregonstate.edu/files/final_report_agroforestry_synthesis_paper_3_14_2013.pdf

Boissière M, Locatelli B, Sheil D, Padmanaba M, Sadjudin E (2013) Local perceptions of climate variability and change in tropical forests of Papua, Indonesia. Ecol Soc 18(4):13

Boko M, Niang I, Nyong A, Vogel C, Githeko A, Medany M, Osman-Elasha B, Tabo R, Yanda P (2007) Africa. Climate change 2007: impacts, adaptation and vulnerability. Contribution of Working Group II to the fourth assessment report of the Intergovernmental Panel on Climate Change, Parry ML, Canziani OF, Palutikof JP, van der Linden PJ, Hanson CE (eds). Cambridge University Press, Cambridge, UK, pp 433–467. https://www.ipcc.ch/pdf/assessment-report/ar4/wg2/ar4-wg2-chapter9.pdf

Bryan E, Deressa TT, Gbetibouo GA, Ringler C (2009) Adaptation to climate change in Ethiopia and South Africa: options and constraints. Environ Sci Policy 12:413–426. https://www.researchgate.net/publication/222015072.pdf

Buishand TA, Brandsma T (1999) Dependence of precipitation on temperature at Florence and Livorno (Italy). Clim Res 12:53–63

CDCCP Synthesis Report (2009) Coping with drought and climate change project: case of the Chiredzi district of Zimbabwe. http://adaptation-undp.org/sites/default/files/downloads/cwd_va_synthesis_report.pdf

Challinor AJ (2009) Developing adaptation options using climate and crop yield forecasting at seasonal to multi-decadal timescales. Environ Sci Policy 12(4):453–465. ISSN 0168-1923

Challinor AJ, Wheeler TR (2008) Crop yield reduction in the tropics under climate change: processes and uncertainties. Agric For Meteorol 148:343–356

Chanana-Nag N, Aggarwal PK (2018) Woman in agriculture, and climate risks: hotspots for development. Clim Chang. https://doi.org/10.1007/s10584-018-2233-z

Chen Y-L, Wang J-J (1995) The effects of precipitation on the surface temperature and airflow over the island of Hawaii. Am Meteorol Soc 123:681–694

Cong R-G, Brady M (2012) The interdependence between rainfall and temperature: copula analyses. Sci World J 2012:405675, 11 pages. https://doi.org/10.1100/2012/405675

De Wit M (2006) Measuring the economic impacts of climate change on crop farming in Africa. Climate change and African agriculture policy note no 8. CEEPA, University of Pretoria, Pretoria

Deressa T, Hassan RM, Alemu T, Yesuf M, Ringler C (2008) Analyzing the determinants of farmers' choice of adaptation methods and perceptions of climate change in the Nile Basin of Ethiopia. IFPRI discussion paper 00798. International Food Policy Research Institute, Washington, DC, 26 p. https://core.ac.uk/download/pdf/6337745.pdf

Deressa TT, Hassan RM, Ringler C, Alemu T, Yesuf M (2009) Determinants of farmers' choice of adaptation methods to climate change in the Nile Basin of Ethiopia. Glob Environ Chang 19:248–255. http://www.sciencedirect.com/science/article/pii/S0959378009000156.pdf

Deressa TT, Ringler C, Hassan RM (2010) Factors affecting the choices of coping strategies for climate extremes: the case of farmers in the Nile Basin of Ethiopia. IFPRI discussion paper no. 01032. International Food Policy Research Institute, Washington, DC, 25 pp. http://ebrary.ifpri.org/cdm/ref/collection/p15738coll2/id/5198.pdf

Deressa TT, Hassan RM, Ringler C (2011) Perception of and adaptation to climate change by farmers in the Nile Basin of Ethiopia. J Agric Sci 149:23–31

Di Falco S, Veronesi M, Yesuf M (2011) Does adaptation to climate change provide food security? A micro-perspective from Ethiopia, am. J Agric Econ 93(3):829–846. https://doi.org/10.1093/ajae/aar006

Easterling WE, Aggarwal PK, Batima P, Brander KM, Erda L, Howden SMA, Kirilenko A, Morton J, Soussana JF, Schmidhuber J, Tubiello FN (2007) Food, fibre and forest products. Climate Change 2007: Impacts, adaptation and vulnerability. Contribution of Working Group II to the fourth assessment report of the Intergovernmental Panel on Climate Change, Parry ML, Canziani OF, Palutikof JP, van der Linden PJ, Hanson CE (eds). Cambridge University Press, Cambridge, UK, pp 273–313. https://www.ipcc.ch/pdf/assessment-report/ar4/wg2/ar4-wg2-chapter5.pdf

FAO, IFAD, UNICEF, WFP, WHO (2018) The state of food security and nutrition in the world 2018. Building climate resilience for food security and nutrition. FAO, Rome. Licence: CC BY-NC-SA 3.0 IGO

Folke C, Carpenter SR, Walker B, Scheffer M, Chapin T, Rockström J (2010) Resilience thinking: integrating resilience, adaptability and transformability. Ecol Soc 15(4):20. [Online]. URL: http://www.ecologyandsociety.org/vol15/iss4/art20/.pdf

Food and Agriculture Organization (FAO) (2006) Livelihood Adaptation to climate variability and change in drought-prone areas of Bangladesh, Rome, p 97. ftp://ftp.fao.org/docrep/fao/009/a0820e/a0820e.pdf

Food and Agriculture Organization (FAO) (2008) Climate change adaptation and mitigation in the food and agriculture sector. High Level Conference on World Food Security – background paper HLC/08/BAK/1. ftp://ftp.fao.org/docrep/fao/meeting/013/ai782e.pdf

Food and Agriculture Organization (FAO) (2009a) Profile for climate change, Rome, 28 pp. ftp://ftp.fao.org/docrep/fao/012/i1323e/i1323e00.pdf

Food and Agriculture Organization (FAO) (2009b) Climate change and agriculture policies; How to mainstream climate change adaptation and mitigation into agriculture policies, Rome, 76 pp. http://www.fao.org/fileadmin/templates/ex_act/pdf/Climate_change_and_agriculture_policies_EN.pdf

Food and Agriculture Organization (FAO) (2010a) Homestead gardens in Bangladesh. Technology for agriculture. Proven technologies for small holders. http://www.fao.org/teca/content/homestead-gardens-bangladesh.pdf. Accessed 26 October 2014

Food and Agriculture Organization (FAO) (2010b) Incorporation of tree management into land management in Jamaica – guinea grass mulching. Technology for agriculture. Proven technologies for small holders. http://www.fao.org/teca/content/incorporation-tree-management-land-management-jamaica-%C2%BF-guinea-grass-mulching.pdf

Food and Agriculture Organization (FAO) (2010c) Collaborative change; a communication framework for climate change adaptation and food security, Rome, 47 pp. http://www.fao.org/docrep/012/i1533e/i1533e00.pdf

Food and Agriculture Organization (FAO) (2011) Framework programme on climate change adaptation, Fao-Adapt. http://www.fao.org/docrep/014/i2316e/i2316e00.pdf

Food and Agriculture Organization (FAO) (2016) Climate change and food security: risks and responses. http://www.fao.org/3/a-i5188e.pdf

Gbetibouo AG (2009) Understanding farmers' perceptions and adaptations to climate change and variability: the case of the Limpopo Basin, South Africa. IFPRI discussion paper no. 00849. International Food Policy Research Institute, Washington, DC, 36 pp. http://www.ifpri.org/publication/understanding-farmers-perceptions-and-adaptations-climate-change-and-variability.pdf

Gbetibouo GA, Ringler C (2009) Mapping South African farming sector vulnerability to climate change and variability; a sub national assessment. IFPRI discussion paper 00885. http://www.ifpri.org/publication/mapping-south-african-farming-sector-vulnerability-climate-change-and-variability.pdf

Gordon CR (2009) The science of climate change in Africa: impacts and adaptation, Grantham Institute for Climate Change, discussion paper N^O 1. Imperial College London, https://workspace.imperial.ac.uk/climatechange/public/pdfs/discussion_papers/Grantham_Institue_-_The_science_of_climate_change_in_Africa.pdf

Hadgu G, Tesfaye K, Mamo G, Kassa B (2015) Farmers' climate change adaptation options and their determinants in the Tigray region, northern Ethiopia. Afr J Agric Res 10(9):956–964. http://www.academicjournals.org/journal/AJAR/article-full-text/9A1563251045.pdf

Hassan R, Nhemachena C (2008) Determinants of African farmers' strategies for adapting to climate change: multinomial choice analysis. Afr J Agric Resour Econ 2(1):83–104. http://ageconsearch.umn.edu/bitstream/56969/2/0201%20Nhemachena%20%26%20Hassan%20-%2026%20may.pdf

Herrero M, Ringler C, van de Steeg J, Thornton P, Zhu T, Bryan E, Omolo A, Koo J, Notenbaert A (2010) Climate variability and climate change: impacts on Kenyan agriculture. International Food Policy Research Institute, Washington, DC. https://cgspace.cgiar.org/bitstream/handle/10568/3840/climateVariability.pdf

Innocent NM, Bitondo D, Azibo BR (2016) Climate variability and change in the Bamenda highlands of the north-west region of Cameroon: perceptions, impacts and coping mechanisms. Br J Appl Sci Technol 12(5):1–18

Intergovernmental Panel on Climate Change (IPCC) (2007) Climate change 2007: impacts, adaptation and vulnerability. Summary for policymakers, IPCC AR4 WGII. Cambridge University Press, Cambridge, UK. https://www.ipcc.ch/pdf/assessment-report/ar4/wg2/ar4-wg2-spm.pdf

International Fund for Agricultural Development (IFAD) (2012) Sustainable small-scale agriculture: feeding the world, protecting the planet. Proceedings of the Governing Council Events. In conjunction with the thirty-fifth session of IFAD's Governing Council, February 2012. https://www.ifad.org/documents/10180/6d13a7a0-8c57-42ec-9b01-856f0e994054.pdf

IPCC, (2014). Climate change 2014: synthesis report. Contribution of working groups I, II and III to the fifth assessment report of the Intergovernmental Panel on Climate Change [Core Writing Team, Pachauri RK, Meyer LA (eds)]. IPCC, Geneva, 151 pp. https://www.ipcc.ch/pdf/assessment-report/ar5/syr/SYR_AR5_FINAL_full.pdf

IPCC (2018) Summary for policymakers. In: Masson-Delmotte V, Zhai P, Pörtner HO, Roberts D, Skea J, Shukla PR, Pirani A, Moufouma-Okia W, Péan C, Pidcock R, Connors S, Matthews JBR, Chen Y, Zhou X, Gomis MI, Lonnoy E, Maycock T, Tignor M, Waterfield T (eds) Global warming of 1.5°C. An IPCC special report on the impacts of global warming of 1.5°C above pre-industrial levels and related global greenhouse gas emission pathways, in the context of strengthening the global response to the threat of climate change, sustainable development, and efforts to eradicate poverty. World Meteorological Organization, Geneva, 32 pp

Ishaya S, Abaje IB (2008) Indigenous people's perception on climate change and adaptation strategies in Jema'a local government area of Kaduna state. Nigeria J Geogr Reg Plann 1(8):138–143

Juana JS, Zibanani K, Okurut FN (2013) Farmers' perceptions and adaptations to climate change in sub-Sahara Africa: a synthesis of empirical studies and implications for public policy in African agriculture. J Agric Sci 5:121–135

Kabir KH, Billah MM, Sarker MA, Miah MAM (2015) Adaptation of farming practices by small-scale farmers in response to climate change. J Agric Ext Rural Dev 7(2):33–40. http://www.academicjournals.org/journal/JAERD/article-full-text/EB533A551012

Kurukulasuriya P, Mendelsohn R (2007) Crop selection: adapting to climate change in Africa. World Bank policy research working paper 4307, Sustainable Rural and Urban Development Team. http://library1.nida.ac.th/worldbankf/fulltext/wps04307.pdf

Kuwornu MKJ, Hassan MR, Etwire MP, Osei-Owusu Y (2013) Determinants of choice of indigenous climate related strategies by small-scale farmers in northern Ghana. British Journal of Environment and Climate Change 3(2):172–187. http://www.sciencedomain.org/review-history.php?iid=251&id=10&aid=1785.pdf

Maddison D (2006) The perception of and adaptation to climate change in Africa. Centre for Applied Environmental Economics and Policy in Africa (CEEPA). Discussion paper no 10. CEEPA, University of Pretoria, Pretoria. http://www.ceepa.co.za/uploads/files/CDP10.pdf

Maddison D (2007) Perception and adaptation to climate change in Africa. Policy research working paper, 4308

Majule AE, Ngongondo C, Kallanda-Sabola M, Lamboll R, Stathers T, Liwenga E, Ngana OJ (2008) Strengthening local agricultural innovation systems in less and more favoured areas of Tanzania and Malawi to adapt to climate change and variability: perceptions. Impacts, vulnerability and adaptation. Res. Abstracts. Issue Number 3: Sokoine University of Agric.: ISBN: 9987-38-9

Mary AL, Majule AE (2009) Impacts of climate change, variability and adaptation strategies on agriculture in semi arid areas of Tanzania: the case of Manyoni District in Singida region, Tanzania. Afr J Environ Sci Technol 3(8):206–218

Mbow C, Smith P, Skole D, Duguma L, Bustamante M (2013) Achieving mitigation and adaptation to climate change through sustainable agroforestry practices in Africa. Curr Opin Environ Sustain 6:8–14. 2014. http://ac.els-cdn.com/S1877343513001255/1-s2.0-S1877343513001255-main. pdf?_tid=84bce8a4-ea32-11e5-ab6e-00000aab0f6b&acdnat=1457994020_73ea11926686e4ef3ccb03abe1829c42

Mbow C, Noordwijk MV, Luedeling E, Neufeldt H, Minang PA, Kowero G (2014) Agroforestry solutions to address food security and climate change challenges in Africa. Curr Opin Environ Sustain 6:61–67. http://ac.els-cdn.com/S1877343513001449/1-s2.0-S1877343513001449-main.pdf?_tid=45e8adc0-ea32-11e5-9b7f-00000aacb35f&acdnat=1457993914_1fcfbc75960e7afcce80d0f1082f300e

McCarthy J, Martello J, Marybeth L (2004) Climate change in the context of multiple stressors and resilience. ACIA scientific report. Cambridge University Press, pp 945–983. www.acia.uaf.edu/PDFs/ACIA_Science_Chapters_Final/ACIA_Ch17_Final.pdf

Mersha AA, Laerhoven FV (2016) A gender approach to understanding the differentiated impact of barriers to adaptation: responses to climate change in rural Ethiopia. Reg Environ Chang. https://doi.org/10.1007/s10113-015-0921-z

Mertz O, Mbow C, Reenberg A, Diouf A (2009) Farmers' perceptions of climate change and agricultural adaptation strategies in rural Sahel. Environ Manag 43(5):804–816

Molua EL (2006) Climate trends in Cameroon: implications for agricultural management. Clim Res 30:255–262

Molua EL (2008) Turning up the heat on African agriculture: the impact of climate change on Cameroon's agriculture. AfJARE 2(1):–20. http://ageconsearch.umn.edu/bitstream/56967/2/0201%20si%20malua%20-%2026%20may.pdf

Molua EL, Lambi CM (2006) The economic impact of climate change on agriculture in Cameroon. In CEEPA discussion paper no. 17

Mortimore MJ, Adams WM (2001) Farmer adaptation, change and crisis in the Sahel. Glob Environ Chang 11:49–57

Mtambanengwe F, Mapfumo P, Chikowo R, Chamboko T (2012) Climate change and variability: small-scale farming communities in Zimbabwe portray a varied understanding. African Crop Science Journal 20(Suppl 2):227–241. http://www.bioline.org.br/pdf?cs12041pdf

Mulenga BP, Wineman A (2014) Climate trends and farmers' perceptions of climate change in Zambia. Working paper no. 86 September 2014. Indaba Agricultural Policy Research Institute (IAPRI) Lusaka, Zambia. http://fsg.afre.msu.edu/zambia/wp86_rev.pdf

National Academy of Sciences (NAS) and the Royal Society (RS) (2014) Climate change, evidence and causes. An overview from the Royal Society and an the United States Academy of Sciences. http://dels.nas.edu/resources/static-assets/exec-office-other/climate-change-full.pdf

Nhemachena C, Hassan R (2007) Micro-level analysis of farmers' adaptation to climate change in southern Africa. IFPRI discussion paper no. 00714. International Food Policy Research Institute, Washington, DC, 30 pp. https://ipcc-wg2.gov/njlite_download2.php?id=8035pdf

Niles MT, Salerno JD (2018) A cross-country analysis of climate shocks and small-scale food insecurity. PLoS One 13(2):e0192928. https://doi.org/10.1371/journal.pone.0192928

Nkuna TR, Odiyo JO (2016) The relationship between temperature and rainfall variability in the Levubu sub-catchment, South Africa. Int J Environ Sci 1:66–75. http://iaras.org/iaras/journals/ijes

Nyanga PH, Johnsen FH, Aune JB, Kalinda TH (2011) Small-scale farmers' perceptions of climate change and conservation agriculture: evidence from Zambia. J Sustain Dev 4(4):73–85

Nzeadibe TC, Egbule CL, Chukwuone NA, Agwu AE, Agu VC (2012) Indigenous innovations for climate change adaptation in the Niger Delta region of Nigeria. Environ Dev Sustain 14(6):901–914

Olsson T, Jakkila J, Veijalainen N, Backman L, Kaurola J, Vehviläinen B (2015) Impacts of climate change on temperature, precipitation and hydrology in Finland – studies using bias corrected Regional Climate Model data. Hydrol Earth Syst Sci 19:3217–3238. https://doi.org/10.5194/hess-19-3217-2015

Rao KPC, Ndegwa WG, Kizito K, Oyoo A (2011) Climate variability and change: farmer perceptions and understanding of intra-seasonal variability in rainfall and associated risk in semi-arid Kenya. Exp Agric 47:267–291. https://doi.org/10.1017/S0014479710000918

Rurinda JP (2014) Vulnerability and adaptation to climate variability and change in small-scale farming systems in Zimbabwe. Thesis submitted in fulfillment of the requirements for the Degree of Doctor at Wageningen University, and publicly defended on Tuesday 10 June 2014 at 11 a.m. in the Aula. http://edepot.wur.nl/305159pdf

Rurinda JP, Mapfumo P, Van Wijk MT, Mtambanengwe F, Rufino MC, Chikowo R, Giller KE (2014) Sources of vulnerability to a variable and changing climate among small-scale households in Zimbabwe: a participatory analysis. Clim Risk Manag 3:65–78

Seleshi Y, Zanke U (2004) Recent changes in rainfall and rainy days in Ethiopia. Int J Climatol 24:973–983

Tabi FO, Adiku SGK, Kwadwo O, Nhamo N, Omoko M, Atika E, Mayebi A (2012) Perceptions of rain-fed lowland rice farmers on climate change, their vulnerability, and adaptation strategies in the Volta region of Ghana. Technol Innov Dev. https://doi.org/10.1007/978-2-8178-0268-8_12

Tambo JA, Abdoulaye T (2012) Climate change and agricultural technology adoption: the case of drought tolerant maize in rural Nigeria. Mitig Adapt Strateg Glob Chang 17:277–292. http://link.springer.com/article/10.1007%2Fs11027-011-9325-7htlm

Temesgen D, Yehualashet H, Rajan DS (2014) Climate change adaptation of small-scale farmers in south eastern Ethiopia. J Agric Ext Rural Dev 6(11):354–366

Tessema YA, Aweke CS, Endris GS (2013) Understanding the process of adaptation to climate change by small-scale farmers: the case of East Hararghe zone, Ethiopia. Agric Food Econ 1:13. http://agrifoodecon.springeropen.com/articles/10.1186/2193-7532-1-13

The Royal Society (2010). Climate change: a summary of the science, September 2010. https://royalsociety.org/~/media/Royal_Society_Content/policy/publications/2010/4294972962.pdf

Thorlakson T (2011) Reducing subsistence farmers' vulnerability to climate change: the potential contributions of agroforestry in western Kenya, occasional paper 16. World Agroforestry Centre, Nairobi. http://www.worldagroforestry.org/downloads/Publications/PDFS/OP11183.pdf

Tingem M, Rivington M, Bellocchi G (2009) Adaptation assessments for crop production in response to climate change in Cameroon. Agron Sustain Dev 29(2):247–256. Springer Verlag (Germany), 2009. http://link.springer.com/article/10.1051%2Fagro%3A2008053

Weng F, Zhang W, Wu X, Xu X, Ding Y, Li G, Liu Z, Wang S (2017) Impact of low-temperature, overcast and rainy weather during the reproductive growth stage on lodging resistance of rice. Sci Rep 7:1–9

World Bank (2013) Turn down the heat: climate extremes, regional impacts, and the case for resilience. A report for the World Bank by the Potsdam Institute for Climate Impact Research and Climate Analytics. World Bank, Washington, DC. License: Creative Commons Attribution – Non Commercial–NoDerivatives3.0 Unported license (CC BY-NC-ND 3.0). http://www.worldbank.org/content/dam/Worldbank/document/Full_Report_Vol_2_Turn_Down_The_Heat_%20Climate_Extremes_Regional_Impacts_Case_for_Resilience_Print%20version_FINAL.pdf

Yaro JW (2013) The perception of and adaptation to climate variability/change in Ghana by small-scale and commercial farmers. Reg Environ Chang 13(6):1259–1272

Zivanomoyo J, Mukarati J (2012) Determinants of choice of crop variety as climate change adaptation option in arid regions of Zimbabwe. Russ J Agric Socio-Econ Sci 3(15). http://www.rjoas.com/issue-2013-03/i015_article_2013_08.pdf

PERMISSIONS

LIST OF CONTRIBUTORS

Wilfred A. Abia
Laboratory of Pharmacology and Toxicology, Department of Biochemistry, Faculty of Science, University of Yaounde 1, Yaounde, Cameroon
School of Agriculture, Environmental Sciences, and Risk Assessment, College of Science, Engineering and Technology (COSET), Institute for Management and Professional Training (IMPT), Yaounde, Cameroon
Integrated Health for All Foundation (IHAF), Yaounde, Cameroon

Comfort A. Onya
Natural Resources and Environmental Management, University of Buea, Buea, Cameroon

Conalius E. Shum and Williette E. Amba
School of Agriculture, Environmental Sciences, and Risk Assessment, College of Science, Engineering and Technology (COSET), Institute for Management and Professional Training (IMPT), Yaounde, Cameroon

Kareen L. Niba
School of Agriculture, Environmental Sciences, and Risk Assessment, College of Science, Engineering and Technology (COSET), Institute for Management and Professional Training (IMPT), Yaounde, Cameroon
Integrated Health for All Foundation (IHAF), Yaounde, Cameroon

Eucharia A. Abia
Integrated Health for All Foundation (IHAF), Yaounde, Cameroon

Floney P. Kawaye and Michael F. Hutchinson
Fenner School of Environment and Society, Australian National University, Canberra, ACT, Australia

Newton R. Matandirotya, Dirk P. Cilliers, Roelof P. Burger and Stuart J. Piketh
Unit for Environmental Sciences and Management, North-West University, Potchefstroom, South Africa

Christian Pauw
NOVA Institute, Pretoria, South Africa

Izael da Silva
Strathmore University, Nairobi, Kenya

Daniele Bricca
Sapienza University of Rome, Rome, Italy

Andrea Micangeli
DIMA, Sapienza University of Rome, Rome, Italy

Davide Fioriti and Paolo Cherubini
DESTEC, University of Pisa, Pisa, Italy

Shilpa Muliyil Asokan
Climate Change and Sustainable Development, The Nordic Africa Institute, Uppsala, Sweden

Joy Obando
Department of Geography, Kenyatta University, Nairobi, Kenya

Brian Felix Kwena
Kenya Water for Health Organization, Nairobi, Kenya

Cush Ngonzo Luwesi
University of Kwango, Kenge, Democratic Republic of Congo

Olukunle Olaonipekun Oladapo
Department of Science Laboratory Technology, Ladoke Akintola University of Technology, Ogbomoso, Nigeria

Leonard Kofitse Amekudzi and Marian Amoakowaah Osei
Department of Physics, Kwame Nkrumah University of Science and Technology, Kumasi, Ghana

Olatunde Micheal Oni
Department of Pure and Applied Physics, Ladoke Akintola University of Technology, Ogbomoso, Nigeria

Abraham Adewale Aremu
Department of Physics with Electronics, Dominion University, Ibadan, Oyo, Nigeria

A. S. Momodu
Centre for Energy Research and Development, Obafemi Awolowo University, Ile-Ife, Nigeria

E. F. Aransiola, T. D. Adepoju and I. D. Okunade
Department of Chemical Engineering, Obafemi Awolowo University, Ile-Ife, Nigeria

Hupenyu A. Mupambwa
Desert and Coastal Agriculture Research, Sam Nujoma Marine and Coastal Resources Research Centre (SANUMARC), Sam Nujoma Campus, University of Namibia, Henties Bay, Namibia

Martha K. Hausiku
Mushroom Research, SANUMARC, Sam Nujoma Campus, University of Namibia, Henties Bay, Namibia

Andreas S. Namwoonde
Renewable Energy Research, SANUMARC, Sam Nujoma Campus, University of Namibia, Henties Bay, Namibia

Gadaffi M. Liswaniso
Mariculture Research, SANUMARC, Sam Nujoma Campus, University of Namibia, Henties Bay, Namibia

Mayday Haulofu
Water Quality Research, SANUMARC, Sam Nujoma Campus, University of Namibia, Henties Bay, Namibia

Samuel K. Mafwila
Oceanography Research, SANUMARC, Sam Nujoma Campus, University of Namibia, Henties Bay, Namibia
Department of Fisheries and Aquatic Science, Sam Nujoma Campus, University of Namibia, Henties Bay, Namibia

Mabvuto Mwanza
Department of Electrical and Electronic Engineering, School of Engineering, University of Zambia, Lusaka, Zambia

Koray Ulgen
Ege University, Solar Energy Institute, Bornova/Izmir, Turkey

Tafadzwa Mutambisi
Department of Demography Settlement and Development, University of Zimbabwe, Harare, Zimbabwe

Nelson Chanza
Department of Geography, Bindura University of Science Education, Bindura, Zimbabwe

Abraham R. Matamanda
Department of Urban and Regional Planning, University of the Free State, Bloemfontein, South Africa

Roseline Ncube
Faculty of Gender and Women Studies, Women's University in Africa, Harare, Zimbabwe

Innocent Chirisa
Department of Demography Settlement and Development, University of Zimbabwe, Harare, Zimbabwe
Department of Urban and Regional Planning, University of the Free State, Bloemfontein, South Africa

Nyong Princely Awazi, Martin Ngankam Tchamba, Lucie Felicite Temgoua and Marie-Louise Tientcheu-Avana
Department of Forestry, Faculty of Agronomy and Agricultural Sciences, University of Dschang, Dschang, Cameroon

Index

A

Adaptive Capacity, 14, 63, 79, 81, 199, 207-209, 213-218, 220-229

Aging, 47

Anthropogenic Activities, 4, 9-10, 127, 140, 209

B

Biodiversity, 5, 10-11, 80, 124, 141, 196, 199, 206

C

Carbon Dioxide, 16-17, 19, 31, 33, 127-128, 194, 209

Cassava, 14-17, 19-20, 22-33

Climate Change, 1-12, 14-15, 17, 19, 31-34, 36-38, 49-56, 59-61, 66, 68-74, 76-89, 97, 99-101, 103-106, 110, 118-127, 129-130, 133-134, 137, 140-143, 145, 192-234

Climate Hazards, 10

Climatic Data, 93

Coastal Lowlands, 2-4, 7, 9

Conservation Agriculture, 137, 140, 233

Crop Failure, 211-212

Crop Productivity, 9

Crop Yields, 11, 15, 17-18, 25, 31, 212

D

Deforestation, 7, 64, 127, 151, 193, 196, 200, 205, 209

Desertification, 5, 151

Disease Control, 50

Droughts, 5-6, 8, 12, 25, 33, 55, 61, 64, 67, 78-82, 84, 130, 198, 206, 211

Dry Season, 68-69

E

Ecosystem, 2-3, 6, 8-9, 13, 15, 59, 69, 82-83, 126, 128, 140, 142, 150-151, 162, 192-193, 196, 198-199, 203-205

Environmental Health, 2, 9-10

Environmental Temperature, 137

F

Farming Practices, 213, 226, 232

Farming Systems, 16, 209, 212, 229, 234

Floods, 5, 7, 12, 33, 38, 55-56, 61, 64, 67, 78-80, 82, 84, 125, 151, 198, 203, 211

Food Crop, 14-16, 124

Food Insecurity, 11, 55, 63, 79, 202, 212, 229, 233

Food Security, 1-3, 5, 8-10, 14-15, 20, 54, 61, 64, 70, 73, 79, 82, 84-85, 136, 201, 206, 230-231, 233

G

Germplasm, 32

Global Warming, 4-5, 37, 50-52, 54, 74, 79, 85, 90, 140, 194, 209, 232

Greenhouse Gas, 4, 51, 85, 97, 101, 104, 118, 120, 125, 145, 151, 194, 232

Gross Domestic Product, 3, 60

H

Heat Stress, 51, 212

Heatwaves, 12, 37-38

Heavy Rainfall, 6, 92

Humidity, 4, 49, 91-93, 134, 142, 150

Hybrid System, 68, 195

I

Industrial Farming, 9

Intercropping, 32, 138, 142

L

Land Degradation, 151

Legumes, 67

Livestock, 64, 74, 117, 124, 130, 211, 213, 215, 220, 222

M

Maize Crop, 24

Methane, 113, 209

Minerals, 133, 135-136, 148

N

Natural Disasters, 2, 6, 8-9, 78, 198

Nutrition, 73, 231

O

Organic Agriculture, 123, 136-138, 141
Organic Food, 141
Organic Matter, 136-137, 139
Oxygen, 124

P

Pesticides, 137
Phytochemicals, 129
Precipitation, 4, 32, 89, 91, 93-94, 97-98, 100-102, 125-126, 134, 211-212, 229-230, 234
Precision, 25, 32
Preliminary Study, 142
Proteins, 133
Public Health, 2, 9, 50-52, 74, 119-120

R

Rainwater Harvesting, 83
Rainy Season, 16, 92, 94
Rationale, 153
Renaissance, 206
Renewable Energy, 55, 61, 66, 72-73, 104-106, 119-122, 131, 145-146, 149, 157, 189-191
Rural Population, 63, 80, 105

S

Saltwater Intrusion, 2-3, 6, 9, 79
Sea Level, 1-5, 7-9, 11-12, 35, 203, 206
Semiarid Lands, 79
Small-scale Farmers, 198, 207-229, 232-234
Smallholder Farmers, 137, 205, 229
Sustainable Agriculture, 2
Sustainable Development, 51, 57, 60, 68, 72-73, 76-79, 81, 83, 85, 104, 119, 186, 188, 191, 195-197, 203-204, 209, 232

T

Thermoregulation, 37, 47
Topography, 12, 89-90, 92, 150-151, 217, 221
Transmission, 50, 56, 65, 146, 151-152, 155, 163, 197
Tree Planting, 64

W

Water Conservation, 84
Water Infiltration, 136
Water Resources, 68, 76-77, 79-84, 86, 88, 124, 129-131, 142
Weather Patterns, 5, 124-125, 210
Wet Season, 16-17, 23, 25, 92

Printed in the USA
CPSIA information can be obtained
at www.ICGtesting.com
JSHW051621061123
51533JS00005B/57

9 781647 403423